STATISTICAL ANALYSIS OF MOLECULAR AND GENOMIC EVOLUTION

Statistical Analysis of Molecular and Genomic Evolution

Xun Gu

Iowa State University

OXFORD
UNIVERSITY PRESS

Great Clarendon Street, Oxford, OX2 6DP,
United Kingdom

Oxford University Press is a department of the University of Oxford.
It furthers the University's objective of excellence in research, scholarship,
and education by publishing worldwide. Oxford is a registered trade mark of
Oxford University Press in the UK and in certain other countries

Published in the United States of America by Oxford University Press
198 Madison Avenue, New York, NY 10016, United States of America

British Library Cataloguing in Publication Data

Data available

Library of Congress Control Number: 2024940577

ISBN 9780198816515
ISBN 9780198816522 (pbk)

DOI: 10.1093/oso/9780198816515.001.0001

Printed and bound by
CPI Group (UK) Ltd, Croydon, CR0 4YY

Links to third party websites are provided by Oxford in good faith and
for information only. Oxford disclaims any responsibility for the materials
contained in any third party website referenced in this work.

Contents

Preface and Acknowledgments

Descending from a single ancestor, genomes became subject to Darwinian selection and adaptation. It is through mutations that functional innovation or loss of ancestral function can lead to an increase of genome complexity, and the adaptation to the challenges of a continuously changing environment. Evolutionary genomics is a recently emerged research field, with ultimate goal of understanding the underlying evolutionary and genetic mechanisms for such processes. It stems from the combinations between high throughput data in functional genomics, statistical modeling and bioinformatics, and phylogeny-based analysis.

During the last decade, high throughput technologies in genomics have generated enormous genome sequences data in many organisms in all ranges of life forms. Indeed, these technological innovations have transformed evolutionary genomes into a rapidly growing field, witnessed by numerous research papers published in the literature. At this stage, these studies are inevitably not only highly heterogeneous in methodology but also highly controversial in biological interpretation. The purpose of this book is to present a tentative framework of statistical theory and methods that are useful in the study of genomic evolution, and illustrate how to use them in real large-scale data analyses. Since nowadays genomic data analyses are almost software-based, our explanation of statistical methods is focused on the biological foundations and model assumptions rather than the details of computational algorithms and practical implementations.

The author wishes to unify the field of evolutionary genomics by building a theoretically inherent framework that can provide a basis to interpret enormous genomic data by the means of evolution. This is a long-term, ambitious goal that cannot be accomplished within the volume of this book. This book discusses various models and methods, including those developed by the author and his collaborators. Some of them were investigated in the process of writing this book and have been published very recently or for the first time here. The criterion for inclusion was given to statistical models based on realistic biological assumptions and statistical methods that are practically useful. Since the laboratory of the author has been engaged in this area of study since the emergence of genome sciences, this book includes a substantial portion of work conducted by his group. However, the author has acknowledged the impossibility to cover all important topics and references that would be beyond his capability, as well as the mission of this book. Indeed, a balance must be struck between personal preference and different scientific views and approaches. For all these reasons, this book should be viewed as an effort from an individual scientist as a step toward the ultimate goal of synthesis.

This book is written for graduate students and researches in the field of evolutionary genetics/genomics and related fields, as well as scientists in mathematics, statistics, or computer sciences who are interested in computational biology and bioinformatics. If this book is used in the classroom, the author encourages instructors to select its contents that mostly fit their teaching purposes. Since this book is intended to introduce statistical models and methods, the author expects readers to have necessary training in statistics and mathematics. In addition, readers should have some basic knowledge of genome sciences and evolutionary biology.

In this book, the author introduces the basics of molecular evolutionary theory and bioinformatics tools so that the readers can comprehend the related issues discussed. The book then covers a range of topics, chapter by chapter. These topics were organized by the model-driven approach, with a few well-selected examples of data analyses. The author acknowledges that there are a huge number of relevant publications in the literature that cannot be all cited here. In particular, this book does not take much space to discuss the widely used data-driven approach that takes advantage of high computational capacity and high-throughput genomic data. Indeed, after developing integrated, comprehensive databases, one can utilize various efficient IT technologies such as clustering or machine learning to conduct extensive data exploration, in an attempt to find biologically meaningful patterns that can be further tested by experimentation or independent datasets. From the author's perspective, the model-driven approach and the data-driven approach are complementary; both are indispensable for gaining a deep understanding of some fundamental biological problems. Hopefully these explorations will benefit researches in this field. Finally, it should be noted that most of the statistical methods discussed in this book have computer programs available that were developed when they were published.

The author has many people to thank. Jiazhen Tan (C. C. Tan), the founder of modern Chinese genetics, has profoundly influenced my career in genetics and evolution since I was an undergraduate student in Fudan University, China. Zhudong Liu, my master degree supervisor at Fudan University, provided me with a unique opportunity to learn mathematical population genetics. Wen-Hsiung Li, my Ph.D. supervisor at the University of Texas, guided me to the forefront of molecular evolution and phylogeny. Masatoshi Nei, my post-doc supervisor at Penn State University, encouraged me to develop my own research niche as genome science was emerging. I would like to thank all my friends, colleagues, and collaborators during my career development over the past two decades, especially Li Jin, Yunxin Fu, Sudhir Kumar, C-I Wu, Manyuan Long, Dan

Graur, Ziheng Yang, Zhenglong Gu, Gunter Wagner, Duane Enger, Dan Voytas, Tom Peterson, Patrick Schnable, Dan Nettleton, Karin Dorman, Hui-Hsien Chou, Xiaoqiu Huang, Jonathan Wendel, Eric Gaucher, Yaping Zhang, Jun Yu, Ji Yang and Yang Zhong. The author is grateful to former and current students, research associates, and collaborators in his laboratory who have made great contributions to solving many challenging problems in statistical study of genomic evolution, computer program development, and large-scale data analysis, especially to Yufeng Wang, Jianying Gu, Kent Vander Velden, Zhongqi Zhang, Shiquan Wu, Huaijun Zhou, Zhixi Su, Yong Huang, Yanyun Zou, Hongmei Zhang, Xiujuan Wang, Zhixi Su, Yangyun Zou, Zhan Zhou, Zhao Zhang, and Jingwen Yang. The author also extends special acknowledgement to his wife, Wei. Without her fully support and encouragement, this book would not have been possible.

Xun Gu
Ames, Iowa USA
Jiangwan, Shanghai, China
July, 2024

1

Evolutionary Distance Analysis in Molecular Evolution

Molecular evolution is the process of evolution at the level of DNA, RNA, and proteins, and the neutral or nearly neutral evolution model has provided the theoretical basis for its study (Kimura 1968, 1983; Kimura and Ohta 1971; Ohta 1973, 1993; Nei 1987; Li 1997). However, the role of positive selection at the molecular level remains a controversial issue for its study (Gillespie 1991; McDonald and Kreitman 1991; Dean and Golding 1997; Messier and Stewart 1997; Zhang et al. 1998; Bustamante et al. 2000, 2005; Tanenbaum et al. 2005; Nielsen et al. 2007). Recent advances in genomics, including whole-genome sequencing, high-throughput protein characterization, and bioinformatics have led to a dramatic increase in studies in comparative and evolutionary genomics. In this chapter, we concisely introduce some widely-used methods in genomic analysis.

1.1 Evolutionary distance of DNA sequences

Evolutionary distances (d) are fundamental for the study of molecular evolution, which is usually measured by the number of nucleotide or amino acid substitutions per site between two homologous sequences (Li 1997; Nei and Kumar 2000). First, d has been widely used to reconstruct phylogenetic trees of genes and gene families. Second, d is a basic measure for studying the pattern and mechanism of DNA/protein evolution, for example, for testing the molecular clock hypothesis (Wu and Li 1985; Gu and Li 1992; Huang et al. 1997) and for detecting positive selection in sequence evolution (Hughes and Nei 1998). Third, with the assumption of constant rate and reliable fossil records, d can be used to date the divergence time of species (Kumar and Hedges 1998; Hedges and Kumar 2009), or gene/genome duplication events (Wang and Gu 2001; Gu et al. 2002). However, d is generally not equal to the observed number of differences per site

Statistical Analysis of Molecular and Genomic Evolution. Xun Gu, Oxford University Press. © Xun Gu (2024).
DOI: 10.1093/oso/9780198816515.003.0001

between two DNA sequences, because multiple substitutions at a given site may occur, especially when the sequence divergence is large. Therefore, to estimate d, a stochastic (Markov) model for DNA evolution is required. In the following, we discuss these stochastic models and methods.

1.1.1 Jukes and Cantor's model: a tutorial

One of the simplest models of nucleotide substitution is that of Jukes-Cantor (1969). This model assumes that at each site a nucleotide changes to one of the three remaining nucleotides with the same substitution rate (r) per year. Let us now consider two nucleotide sequences, X and Y, which diverged from the common ancestral sequence t years ago. We denote by q_t the probability of identical nucleotides between X and Y and by p_t ($= 1 - q_t$) the probability of different nucleotides. At time $t+1$ (measured in years), the probability of identical nucleotides, q_{t+1}, can be derived as follows (Fig. 1.1).

(*i*) With a probability of q_t, two sequences have the same nucleotide at time t. At time $t+1$, the chance that remains the same is $(1-r)^2 \approx 1 - 2r$, that is, the probability of no change in both lineages. Note that the probability of double changes is negligible.

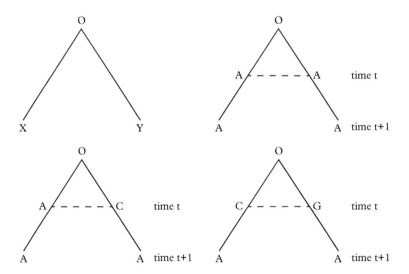

Figure 1.1 *Illustration for the derivation of Jukes-Cantor model in the case of two DNA sequences (X and Y) diverged from the common ancestor (O). For any site that has the same nucleotide (A) in both sequences at the time t+1, at time t the nucleotide could be either the same, or different (one is A and the other is C). Nevertheless, double changes occurred in both lineages in the time interval between t and t+1 are negligible.*

(*ii*) With a probability of $1 - q_t$, two sequences have different nucleotides at time t, the chance at time $t + 1$ that the same nucleotide remains is $2 \times (1/3) \times r$; the factor $1/3$ is for the change to a specific nucleotide, and the factor of 2 indicates that it could happen in either of two lineages. Double changes from t to $t + 1$ are negligible. Together with (*i*), we then have $q_{t+1} = (1 - 2r)q_t + \frac{2}{3}r(1 - q_t)$, which can be written as $q_{t+1} - q_t = \frac{2r}{3} - \frac{8r}{3}q_t$.

(*iii*) Let us now use a continuous time model and represent $q_{t+1} - q_t$ by dq/dt. We then have the following differential equation

$$\frac{dq}{dt} = \frac{2r}{3} - \frac{8r}{3}q_t \qquad (1.1)$$

Solution of this equation with the initial condition $q = 1$ at $t = 0$ gives

$$q = 1 - \frac{3}{4}\left(1 - e^{-8rt/3}\right) \qquad (1.2)$$

(*iv*) Under the Jukes-Cantor (1969) model, the expected number of nucleotide substitutions per site (*d*) for the two sequences is $2rt$. Therefore, *d* can be estimated by

$$\hat{d} = -\frac{3}{4}\ln\left(1 - \frac{4}{3}\hat{p}\right) \qquad (1.3)$$

where $\hat{p}$ is the proportion of different nucleotides between *X* and *Y*.

However, the derivation of the evolutionary distance is tedious when the substitute model becomes sophisticated. In the next section, we introduce a formal mathematical treatment.

1.1.2   Models of nucleotide substitution

The rate matrix of nucleotide evolution

Since at each site in a DNA sequence, there are four possible nucleotides (A, T, C, and G), a model of nucleotide substitution is characterized by a 4×4 rate matrix **R**, which is also called the pattern of nucleotide substitution. This matrix can be represented as shown in Table 1.1: The *ij*-th element of **R**, denoted by r_{ij}, is the substitution rate from nucleotide *i* to nucleotide *j*, $i \neq j$; the diagonal elements are given by $r_{ii} = -\sum_{j \neq i} r_{ij}$, so that the sum of the elements in each row is zero. Thus, the most general model has 12 independent parameters but this model is too complex to apply. Therefore, it is necessary to make some assumptions on **R** to develop a useful method for estimating *d*. As examples, the following are two simple but widely-used models.

Table 1.1 *Models of nucleotide substitution*

(a)

	A	T	C	G
A	$-(a+b+c)$	a	b	c
T	d	$-(d+e+f)$	e	f
C	g	h	$-(g+h+i)$	i
B	j	k	l	$-(j+k+l)$

(b)

	A	T	C	G
A	-3μ	μ	μ	μ
T	μ	-3μ	μ	μ
C	μ	μ	-3μ	μ
B	μ	μ	μ	-3μ

(c)

	A	T	C	G
A	$-(2v+s)$	v	v	s
T	v	$-(2v+s)$	s	v
C	v	s	$-(2v+s)$	v
B	s	v	v	$-(2v+s)$

(d)

	A	T	C	G
A	r_{11}	$\pi_2 v_1$	$\pi_3 v_2$	$\pi_4 s_1$
T	$\pi_1 v_1$	r_{22}	$\pi_3 s_2$	$\pi_4 v_3$
C	$\pi_1 v_2$	$\pi_2 s_2$	r_{33}	$\pi_4 v_4$
B	$\pi_1 s_1$	$\pi_2 v_3$	$\pi_3 v_4$	r_{44}

The first one is Jukes and Cantor's one-parameter model (Jukes-Cantor 1969), in which the substitution rate is assumed to be the same for all nucleotides, i.e., $r_{ij} = u$, for all $i \neq j$, and $r_i i = -3u$; see Table 1.1 for the rate matrix **R**. The second is Kimura's two-parameter model (Kimura 1980), in which the rate of transitional substitution (i.e., changes between A and G or between C and T) may not be equal to that of transversional substitution (i.e., all the other types of changes); the two rates are denoted by s and v, respectively (Table 1.1). This model is more realistic than the one-parameter model because transitions are generally more frequent than transversions in DNA evolution.

The transition probability matrix

In general, let $P_{ij}(t)$ be the transition probability from nucleotide i to j after t time units. By Markovian theory, $P_{ij}(t)$ satisfies the linear differential equation

$$\frac{dP_{ij}(t)}{dt} = \sum_k r_{ik} P_{kj}(t) \tag{1.4}$$

where $i, j, k = A, G, T$, or C, with the initial condition $P_{ii}(0) = 1$ and $P_{ij} = 0 \ (i \neq j)$. Let matrix $\mathbf{P}(t)$ consist of $P_{ij}(t)$. In matrix form, Eq. (1.2) can be expressed as

$$\frac{d\mathbf{P}(t)}{dt} = \mathbf{R}\mathbf{P}(t) \tag{1.5}$$

with the initial condition $\mathbf{P}(0) = \mathbf{I}$, where $\mathbf{I}$ is the identity matrix. The solution of Eq. (1.3) is

$$\mathbf{P}(t) = e^{\mathbf{R}t} \tag{1.6}$$

1.1.3 Evolutionary distance and estimation

Theme of two homologous DNA sequences

Because we generally do not know the initial sequences, to study DNA sequence evolution we need to consider two homologous DNA sequences which diverged from O, the common ancestor, t time units ago (Fig. 1.1). The mean evolutionary rate (per year per site) is given by $\bar{r} = \sum_i f_i \sum_{j \neq i} r_{ij}$, where f_i is the frequency of nucleotide i in the ancestral sequence. Since $\sum_{j \neq i} r_{ij} = -r_{ii}$, it can be simplified as $\bar{r} = -\sum_i f_i r_{ii}$. Thus, the number of substitutions per site between the two sequences, the evolutionary distance (d), is expected to be

$$d = 2\bar{r}t = -2t \sum_i f_i r_{ii} \tag{1.7}$$

where the factor 2 comes from the fact that d is the sum of the numbers of substitutions per site in the two lineages. One problem is that d depends on the initial frequencies f_i, which we do not know. However, if the nucleotide frequencies are stationary, that is, their expectations do not change with time, then the initial frequencies can be estimated by the average frequencies in the present sequences. The assumption of stationarity greatly simplifies the estimation problem.

Stationary and time reversible model

We consider the case where the substitution process is stationary and time reversible (Gu and Li 1996b); this is called the SR model. Time reversibility means that $\mathbf{R}$ is restricted by the following condition:

$$\pi_i r_{ij} = \pi_j r_{ji} \tag{1.8}$$

for any $i \neq j$, where π_i is the equilibrium frequency of nucleotide i; $i = 1, 2, 3, 4$ for A, T, C, and G, and $\pi_1 + \pi_2 + \pi_3 + \pi_4 = 1$. Therefore, the SR model is a nine-parameter model, which includes the one-parameter model, the two-parameter model, and several other models (Tajima and Nei 1982, 1984; Tamura and Nei 1993) as special cases (Table 1.1).

Consider the two-lineage scheme (Fig. 1.1). Since time reversibility implies that the substitution process from the common ancestor O to sequences X and Y is equivalent to that from X through O to Y (or from Y through O to X), the transition probability matrix from X to Y is given by

$$\mathbf{P}(2t) = e^{2t\mathbf{R}} \tag{1.9}$$

By spectral decomposition, the diagonal elements of $\mathbf{R}$ can be expressed as

$$r_{ii} = \sum_{k=1}^{4} u_{ik} v_{ki} \lambda_k \tag{1.10}$$

where λ_k ($k = 1, 2, 3, 4$) is the k-th eigenvalue of $\mathbf{R}$, one of which is zero, say $\lambda_4 = 0$; u_{ik} is the ik-th element of the eigenmatrix $\mathbf{U}$ and v_{ki} is the ki-th element of matrix $\mathbf{V} = \mathbf{U}^{-1}$. Putting Eq. (1.8) into Eq. (1.5) and setting $f_i = \pi_i$, we have

$$d = -2t \sum_{k=1}^{4} b_k \lambda_k \tag{1.11}$$

where the constants $b_k = \sum_{i=1}^{4} \pi_i u_{ik} v_{ki}$; in particular $\lambda_4 = 0$ that means $b_4 = 0$.

In the following, we discuss how to derive the formula for estimating d under a specified substitution model. We will start with the one-parameter and two-parameter models, for which the eigenvalues of $\mathbf{R}$ can be obtained analytically. Then, we shall discuss the nine-parameter SR model, for which the eigenvalues of $\mathbf{R}$ can not be obtained analytically. Besides, we show that all these methods can be extended to the case where the substitution rate varies among nucleotide sites.

One-parameter method

In the one-parameter model, the eigenvalues of $\mathbf{R}$ are given by $\lambda_1 = \lambda_2 = \lambda_3 = -4u$ and $\lambda_4 = 0$ and so $b_1 = b_2 = b_3 = 1/4$. From Eq. (1.9), the number of substitutions per site can be simplified as $d = 6ut$. From the eigenvalues and eigenmatrix of $\mathbf{R}$, the transition probability is given by

$$P_{ij}(t) = \begin{cases} \frac{1}{4} + \frac{3}{4}e^{-4ut} & \text{if } i = j \\ \frac{1}{4} - \frac{1}{4}e^{-4ut} & \text{if } i \neq j \end{cases} \tag{1.12}$$

Now consider the two-lineage scheme (Fig. 1.2). Let $I(t)$ be the probability that, at time t, the nucleotides at a given site in the two sequences are identical to each other. Since the probability that both sequences have nucleotide i is $\sum_{k=1}^{4} f_k P_{kj}^2(t)$, where f_k is the frequency of nucleotide k at the ancestor O, $I(t)$ is given by $I(t) = \sum_{k=1}^{4} f_k \left[P_{k1}^2(t) + P_{k2}^2(t) + P_{k3}^2(t) + P_{k4}^2(t) \right]$. From Eq. (1.10) and after some simplifications, one can show that, regardless of the initial frequencies f_k, $I(t)$ is given by $I(t) = 1/4 + (3/4)e^{-8ut}$. Note that the probability that two sequences are different at a site at time t is $p = 1 - I(t)$, which can be estimated by the observed proportion of different nucleotides between the two sequences. Therefore we have shown that the evolutionary distance $d = 6ut$ can be estimated by Eq. (1.1), that is,

$$d = -\frac{3}{4} \ln \left(1 - \frac{4}{3}p \right)$$

Kimura's two-parameter method

Under Kimura's two parameter model (Kimura 1980), the eigenvalues of $\mathbf{R}$ are $\lambda_1 = \lambda_2 = -2(s+v)$, $\lambda_3 = -4v$ and $\lambda_4 = 0$; the constants are $b_1 = b_2 = 1/2$ and $b_3 = 1/4$. Thus, d is given by $d = 2(s+2v)t$. On the other hand, the transition probability under this model is given by

$$P_{ij}(t) = \begin{cases} \frac{1}{4} + \frac{1}{4}e^{-4vt} + \frac{1}{2}e^{-2(s+v)t} & \text{for } i = j \\ \frac{1}{4} + \frac{1}{4}e^{-4vt} - \frac{1}{2}e^{-2(s+v)t} & \text{for transitions} \\ \frac{1}{4} - \frac{1}{4}e^{-4vt} & \text{for transversions} \end{cases} \tag{1.13}$$

Let P and Q be the probabilities that the two sequences differ by a transition and a transversion, respectively. One can show $P = 2\sum_k f_k (P_{kA}P_{kG} + P_{kT}P_{kC})$ and $P + Q = I(t)$. By Eq. (1.11) one can show

$$P = \frac{1}{4} + \frac{1}{4}e^{-8vt} - \frac{1}{2}e^{-4(s+v)t}$$

$$Q = \frac{1}{2} - \frac{1}{2}e^{-8vt} \tag{1.14}$$

Thus, $d = 2(s + 2v)t$ can be estimated by

$$d = -\frac{1}{2}\ln(1 - 2P - Q) - \frac{1}{4}\ln(1 - 2Q) \tag{1.15}$$

where P and Q can be estimated from the two sequences.

1.1.4   The general stationary and time-reversible model

So far, explicit solutions for estimating d have been obtained up to six-parameter models (Tamura and Nei 1993; Li and Gu 1996). It is difficult to derive an analytical formula for d because the eigenvalues of $\mathbf{R}$ cannot be expressed in an analytical form (Rodríguez et al. 1990). Under the SR (stationary and time-reversible) model, Gu and Li (1996b) provided a practically feasible solution. Let z_k be the k-th eigenvalue of $\mathbf{P}(2t)$. By matrix theory, Eq. (1.7) implies that $\mathbf{R}$ and $\mathbf{P}(2t)$ have the same eigenmatrix, and their eigenvalues have the following relationship

$$z_k = e^{2t\lambda_k} \qquad k = 1,\ldots,4 \tag{1.16}$$

Note that $\lambda_4 = 0$ and $z_4 = 1$. Thus, the evolutionary distance under the SR model can be written as

$$d = -\sum_{k=1}^{3} b_k \ln z_k \tag{1.17}$$

where z_k and b_k can be estimated from sequence data (see below). Under the SR model, all eigenvalues z_k (or λ_k) are real.

To estimate z_k and b_k from sequence data, we must estimate the transition probability matrix $\mathbf{P}(2t)$ first. Let $\mathcal{J}_{ij}$ be the expected frequency of sites where the nucleotide is i in sequence X and j in sequence Y. Let matrix $\mathbf{J}$ consist of $\mathcal{J}_{ij}$. It can be shown that the matrix $\mathbf{J}$ is symmetric under the SR model. By Markovian properties, we have

$$\mathcal{J}_{ij} = \sum_{k=1}^{4} \pi_k P_{ki}(t) P_{kj}(t) \tag{1.18}$$

$i,j = 1,\ldots 4$. By time reversibility, i.e., $\pi_i P_{ij}(t) = \pi_j P_{ji}(t)$, we have

$$\mathcal{J}_{ij} = \sum_{k=1}^{4} \pi_i P_{ik} P_{kj}(t) = \pi_i \sum_{k=1}^{4} P_{ik}(t) P_{kj}(t) = \pi_i P_{ij}(2t) \tag{1.19}$$

where $\sum_{k=1}^{4} P_{ik}(t) P_{kj}(t) = P_{ij}(2t)$ is a basic property of transition probabilities. Thus, Eq. (1.16) gives a simple method for estimating $P_{ij}(t)$ from sequence data $\mathcal{J}_{ij}$:

(a) Count N_{ij}, the number of sites at which the nucleotide is i in sequence X and is j in sequence Y, and then compute $\hat{\mathcal{J}}_{ij} = N_{ij}/N$, where N is the number of nucleotides in the sequence.

(b) Estimate the transition probability matrix $\mathbf{P}(2t)$ by

$$\hat{P}_{ij} = \frac{\hat{\mathcal{J}}_{ij}}{\hat{\pi}_i} \tag{1.20}$$

$(i, j = 1, \ldots, 4)$, where $\hat{\pi}_i$ is the frequency of nucleotide i estimated by taking (simple) average between the two sequences. However, when $i \neq j$, the estimated frequency $\hat{\mathcal{J}}_{ij}$ may not be equal to $\hat{\mathcal{J}}_{ji}$, i.e., the estimated matrix $\hat{\mathbf{J}}$ may not be symmetric. Gu and Li (1996b) proposed a method to test whether $\mathcal{J}_{ij} - \mathcal{J}_{ji}$ $(i \neq j)$ is significantly different from zero. If the null hypothesis is not rejected statistically, the deviation from symmetry can be regarded as sampling effects, and the ij-th and the ji-th elements of $\mathbf{J}$ are equally given by $[\hat{\mathcal{J}}_{ij} + \hat{\mathcal{J}}_{ji}]/2$.

(c) Then, the estimate of z_k $(k = 1 \ldots, 4)$ can be obtained by solving the following characteristic equation

$$\det(\hat{\mathbf{P}} - \hat{z}\mathbf{I}) = 0 \tag{1.21}$$

where $\hat{\mathbf{P}}$ consists of $\hat{P}_{ij}$ and $\mathbf{I}$ is the identity matrix; the corresponding eigenmatrix $\mathbf{U}$ and its inverse matrix $\mathbf{V}$ are also obtained simultaneously by a standard algorithm. Thus, d can be estimated according to Eq. (1.14) and the sampling variance can be approximately computed by the formula given by Gu and Li (1996b).

1.1.5 Estimation of *d* under variable rates

All the above methods for estimating d assume the same substitution rate for all sites. If this assumption is violated, d may be seriously underestimated, especially for divergent sequences. Because the substitution rate usually varies among sites in most genes, these methods need to be modified to take rate variation into account.

There is empirical evidence that the rate variation among sites follows a gamma distribution. This distribution is mathematically simple and is commonly used in the literature. Assume that the ij-th element of the rate matrix $\mathbf{R}$ is expressed by $r_{ij} = h_{ij}u$, where the constant h_{ij} represents the pattern of substitution rate and the random variable u varies among sites according to the following gamma distribution

$$\phi(u) = \frac{\beta^\alpha}{\Gamma(\alpha)} u^{\alpha-1} e^{-\beta u} \tag{1.22}$$

where the mean of u is given by $\bar{u} = \alpha/\beta$.

First, consider the one-parameter model; in this case, $h_{ij} = 1$ for all $i \neq j$. It follows that the mean of p, i.e., the mean proportion of differences between two sequences, is given by

$$\bar{p} = \int_0^\infty \frac{3}{4}\left(1 - e^{-8ut}\right)\phi(u)du = \frac{3}{4}\left[1 - \left(1 + \frac{8\bar{u}t}{\alpha}\right)^{-\alpha}\right] \tag{1.23}$$

Therefore, the average number of substitutions per site $d = 6\bar{u}t$ can be estimated by

$$d = \frac{3}{4}\alpha\left[\left(1 - \frac{4}{3}p\right)^{-1/\alpha} - 1\right] \tag{1.24}$$

In the same manner, Jin and Nei (1990) extended Kimura's two-parameter method (1980) to the case where the rate varies according to a gamma distribution:

$$d = \frac{1}{4}\alpha\left[2\left(1 - 2P - Q\right)^{-1/\alpha} + \left(1 - 2Q\right)^{-1/\alpha} - 3\right] \tag{1.25}$$

Finally, for the SR model, the average d is

$$d = \alpha\sum_{k=1}^{3} b_k\left(z_k^{-1/\alpha} - 1\right) \tag{1.26}$$

where the eigenvalues z_k and the constants b_k can be estimated by the same approach as above for the SR model under the uniform rate model (Gu and Li 1998).

In summary, because the ancestral sequence is generally unknown, the most general model (i.e., 12 parameters) is difficult to apply so that some restrictions on the rate matrix $\mathbf{R}$ are necessary for developing useful methods for estimating d. As reviewed here, many methods have been developed for this purpose. As a general rule, methods based on more general models will have smaller estimation bias but larger sampling variances. Thus, when the sequence length (N) is large, more general methods are preferred. However, when N is small, simpler methods may be better. For example, Gu and Li's (1996b) simulation study suggested that, if N is less than 200 base pairs, the one-parameter method is on average better than the SR method; whereas if N is larger than 500, the SR method is on average better than the one-parameter method.

1.1.6 The LogDet distance

Nucleotide frequency in sequence evolution

The stationarity of nucleotide frequencies is one of the most common assumptions made in estimating evolutionary distances (see Lanave et al. 1984; Zharkikh 1994; Gu and Li

1996b). It assumes that the expectations of nucleotide frequencies in a sequence do not change with time and are equal to those in the ancestral sequence. Therefore, to estimate the distance between two sequences, the nucleotide frequencies in the ancestral sequence are estimated by the averages of the nucleotide frequencies in the two extant sequences. If nucleotide frequencies vary with time so that stationarity does not hold, the estimated distance may not be accurate. As a consequence, a distance-matrix method for phylogeny reconstruction can be misleading, i.e., it tends to group sequences of similar nucleotide frequencies irrespective of the true evolutionary relationships (Hasegawa and Hashimoto 1993; Sogin et al. 1993; Steel 1994).

The LogDet (Lake 1994; Steel 1994; Lockhart et al. 1994; Gu and Li 1996a) distance has been proposed to deal with the nonstationarity problem. In spite of various versions, these methods are based on the most general model of nucleotide substitutions. Historically, these methods can be traced back to Barry and Hartigan (1987) and Cavender and Felsenstein (1987). In this section, we study some statistical properties of the LogDet distance.

The canonical formula

Consider two sequences (denoted by 1 and 2, respectively) that evolved from O, their common ancestor, t time units ago (see Fig. 1.2). Let $\mathbf{J}$ be the data matrix whose ij-th element $\mathcal{J}_{ij}$ is the proportion of sites at which the nucleotide is i in sequence 1 and j in sequence 2. Then, the logDet distance (between sequences 1 and 2) is defined as

$$d = -\frac{1}{4} \ln \det[\mathbf{J}] \qquad (1.27)$$

where *det* means the determinant of a matrix. By using the delta method (Barry and Hartigan 1987), the sampling variance of the estimated LogDet distance $(\hat{d})$ is found to be

$$Var(\hat{d}) \approx \frac{1}{16L} \sum_{i=1}^{4} \sum_{j=1}^{4} (M_{ij}^2 \mathcal{J}_{ij} - 1) \qquad (1.28)$$

where L is the sequence length and M_{ij} is the ij-th element of $\mathbf{M} = \mathbf{J}^{-1}$. The estimation procedure for the LogDet distance is straightforward because the matrix $\mathbf{J}$ can be directly estimated from the sequence data.

Properties

The LogDet distance is potentially very useful in the study of DNA evolution because they have the following nice properties (Gu and Li 1996a):

(1) The LogDet distance is based on the most general model of nucleotide substitution, i.e., the 12-parameter model. Moreover, it is valid even if the rate matrix **R** differs among lineages.

(2) The LogDet distance is useful for phylogenetic reconstruction when nucleotide frequencies are nonstationary. It has been shown (e.g., Gu and Li 1996a) that for some distance matrix methods of phylogenetic reconstruction such as the neighbor-joining (NJ) method (Saitou and Nei 1987), the LogDet distance lead to the correct tree topology when the nucleotide frequencies vary considerably among sequences.

(3) Let μ_1 and μ_2 be the arithmetic mean of the evolutionary rate in lineage 1 or lineage 2, respectively, that is, $\mu_1 = -\sum_{i=1}^{4} r_{ii}^{(1)}/4$ and $\mu_2 = -\sum_{i=1}^{4} r_{ii}^{(2)}/4$. The biological interpretation of the LogDet distance can be described as the following equation

$$d = 2\mu t - \frac{1}{4}\sum_{i=1}^{4} \ln f_{i,0} \tag{1.29}$$

where $\mu = (\mu_1 + \mu_2)/2$ is the mean rate over the two lineages, and $f_{i,0}$ is the frequency of nucleotide i at the ancestral node O. Therefore, the LogDet distance is not only linear in time t, but also depends on the nucleotide frequencies at the ancestral node.

(4) The LogDet distance is useful for testing the molecular clock hypothesis under nonstationary frequencies because the relative-rate test (Wu and Li 1985) is not affected by the ancestral nucleotide frequencies.

On the other hand, the LogDet distance may have several disadvantages that need to be examined carefully in practice:

(1) The sampling variance of LogDet distance becomes large for short sequences. By substantial simulation studies, it is recommended for use when the sequence length is > 500 bp. Gu and Li (1996a) have demonstrated that the LogDet distance is, on average, overestimated especially when the sequences are short; the bias becomes trivial as the sequence length > 2,000 bp. Furthermore, Gu and Li (1996a) proposed an empirical bias-corrected LogDet distance as follows

$$\hat{d}_c = -\frac{1}{4}\ln \det[\hat{J}] - 2Var(\hat{d}) \tag{1.30}$$

Computer simulation showed that the statistical bias can be largely corrected by this formula.

(2) When $t = 0$, $d = -\sum_{i=1}^{4} \ln f_{i,0}/4 > 0$. In other words, the LogDet distance satisfies the non-negative condition but have a non-zero positive constant at the initial condition. One may modify the LogDet distance by

$$d = -\frac{1}{4} \ln \det[\mathbf{J}] - d_0 \qquad (1.31)$$

The problem is that it may cause the violation of the non-negative condition in some cases. Noted that adding $-d_0$ in the LogDet distance has no effect on the performance of tree-making but may affect the branch lengths estimation. In practice, one may calculate the effect of nucleotide frequencies $F = -\sum_{i=1}^{4} \ln f_i/4$ for all extant sequences and choose $d_0 = F_{min}$, the minimum value over all sequences to guarantee the non-negative nature of the evolutionary distance.

(3) Probably the theoretical challenge of the LogDet distance is the assumption of constant rate over nucleotide sites, which is obviously unrealistic. Indeed, when the rate varies among sites, the biological interpretation of the LogDet distance holds only approximately. In the estimation of evolutionary distance, it remains an unsolved problem how to solve the nonstationary problem and the rate variation among sites simultaneously.

1.2 Evolutionary distance between protein-encoding sequences

1.2.1 Poisson distance of protein sequence

In the history of molecular evolution, study of the evolutionary change of proteins began with the comparison of two or more amino acid sequences from different organisms. One simple measure of the extent of sequence divergence is the proportion of different amino acids between two sequences, also called the *p*-distance. However, the proportion of different amino acids (*p*) is not strictly proportional to the divergence time (*t*), as multiple amino acid substitutions start to occur at the same sites.

One simple way to estimate the protein distance more accurately is based on the Poisson process, which claims that the probability for no amino acid substitution to have occurred during t years at a site of a sequence is e^{-vt}, where v is the evolutionary rate. Hence, the probability (q) that neither of the homologous sites of the two protein sequences has undergone substitution is e^{-2vt}, which can be estimated by $q = 1 - p$. It follows that the expected number of amino acid substitutions per site for the two sequences, i.e., the evolutionary distance $d = 2vt$, is given by

$$d = -\ln(1-p) \tag{1.32}$$

It should be noted that the Poisson distance is approximate because backward mutations and parallel mutations (the same mutations occurring at the homologous amino acid sites in two different evolutionary lineages) are not taken into account. However, the effects of these mutations are generally very small unless p is large.

1.2.2   Amino acid substitution matrix

Empirical studies have shown that amino acid substitution occurs more often between amino acids that are similar in biochemical properties than between dissimilar amino acids (Dayhoff 1972). As a result, amino acid substitution is generally not random, and backward and parallel substitutions may occur quite often between similar amino acids. Some amino acids such as cysteine and tryptophan rarely change. To take into account these factors, Dayhoff et al. (1978) proposed the so-called PAM-based method of estimating evolutionary distance. The amino acid substitution matrix for a relatively short period of time is considered, and the relationship between the proportion of identical amino acids and the number of amino acid substitutions is derived empirically.

The amino acid substitution matrix Dayhoff et al. (1978) used was derived from empirical data for many proteins such as hemoglobins, cytochrome c, and fibrinopeptides. They first constructed an evolutionary tree for closely related amino acid sequences and then inferred the relative frequencies of substitutions between different amino acids. From these data, they constructed an empirical amino acid substitution matrix for the 20 amino acids. An element (m_{ij}) of this substitution matrix gives the probability that the amino acid in row i changes to the amino acid in column j during one evolutionary time unit. The time unit used in the matrix is the time during which one amino acid substitution per 100 amino acid sites occurs on average. Dayhoff et al. (1978) measured the number of amino acid substitutions in terms of accepted point mutations (PAM); 1 PAM represents one amino acid substitution per 100 amino acid sites.

Although Dayhoff's substitution matrix is still widely used, Jones et al. (1992) constructed a new matrix based on a large amount of substitution data from many different proteins. Adachi and Hasegawa (1996) also produced a substitution matrix for 13 mitochondrial proteins in vertebrates. Theoretically, different protein families (such as globins or protein kinases) are expected to have different substitution matrices, so it is desirable to construct a substitution matrix for each group of proteins. Nevertheless, it has shown that the evolutionary distance under the empirical substitution matrix can be numerically approximately some simple formulas. For instance, Kimura (1983) shows the following distance

$$d = -\ln(1 - p - 0.2p^2) \tag{1.33}$$

where p is the proportion of different amino acid sites, closely approaches the Dayhoff's distance when $p < 0.8$.

1.2.3 Synonymous and nonsynonymous distances

Many (but not all) nucleotide substitutions at the third position are silent and do not change amino acids. Meanwhile, some silent substitutions may also occur at the first positions. Since synonymous substitutions are most likely free from natural selection (but see Chamary et al. 2006 for different opinions), the rate of synonymous substitution is often equated to the rate of neutral nucleotide substitution. By contrast, the rate of nonsynonymous substitution is generally much lower than that of synonymous substitution and varies extensively from gene to gene. This is considered to be due to purifying selection, the extent of which varies from gene to gene (Kimura 1983). For some genes, on the other hand, nonsynonymous substitutions occur at a higher rate than synonymous substitutions (e.g., Hughes and Nei 1988; Lee et al. 1995). These nonsynonymous substitutions are apparently caused by positive selection, because under neutral evolution one would expect that the rates of synonymous and nonsynonymous substitution are equal to each other. For these reasons, estimation of the rates of synonymous and nonsynonymous substitution has become an important subject in the study of molecular evolution.

Nei-Gojobori method

When the number of nucleotide substitutions between two DNA sequences is so small that there is no more than one nucleotide difference between any pair of homologous codons compared, the numbers of synonymous and nonsynonymous substitutions can be obtained by simply counting silent and amino acid altering nucleotide differences. However, when two or more nucleotide differences exist between a pair of codons, the distinction between synonymous and nonsynonymous substitutions is no longer simple, because of multiple evolutionary pathways between them. Nei and Gojobori (1986) developed an unweighted method to calculate the average numbers of synonymous and nonsynonymous substitutions over multiple evolutionary pathways (Perler et al. 1980; Miyata and Yasunaga 1980).

To estimate the synonymous (d_S) and nonsynonymous distances (d_N), one has to classify the synonymous and nonsynonymous sites as follows: Let i be the number of possible synonymous changes at this site. This is counted as $i/3$ synonymous and $(1 - i/3)$ nonsynonymous. For instance, in the codon TTT (phenylalanine), the first two positions are counted as nonsynonymous sites because no synonymous changes can occur at these positions. Meanwhile, the third position is counted as one-third synonymous and two-third nonsynonymous because one of three possible changes is synonymous. After obtaining the number of synonymous and nonsynonymous sites, it is straightforward to estimate d_S and d_N separately under the Jukes-Cantor model.

Li-Wu-Luo method

Li et al. (1985) proposed an alternative way to the averaging multiple evolutionary pathways, by classifying nucleotide sites into nondegenerate, twofold degenerate, and fourfold degenerate sites. A site is nondegenerate if all possible changes at this site are nonsynonymous, twofold degenerate if one of the three possible changes is synonymous, and fourfold degenerate if all possible changes are synonymous. This method then calculates the number of substitutions between two coding sequences for the three types of sites separately. Note that by definition all the substitutions at nondegenerate sites are nonsynonymous, and all the substitutions at fourfold degenerate sites are synonymous. At twofold degenerate sites, transitional changes (C/T or A/G) are synonymous and other changes (transversions) are nonsynonymous. For two exceptions in the universal genetic code (arginine and isoleucine), Li et al. (1985) suggested an ad hoc correction. Under the Kimura two-parameter model, transitional distance and transversional distance at each type of sites can be calculated. Therefore, the synonymous distance (d_S) is the average (weighted by the number of sites) of the evolutionary distance at the fourfold degenerate sites and the transitional distance at twofold degenerate sites. Similarly, the nonsynonymous distance (d_N) is the average (weighted by the number of sites) of the evolutionary distance at the nondegenerate sites and the transversional distance at the twofold degenerate sites. To correct the bias in the original version of Li et al. (1985), Li (1993) proposed an unbiased method to estimate the rates of synonymous and nonsynonymous substitutions.

Codon substitution models

Goldman and Yang (1994) developed a likelihood method for estimating the rates of synonymous and nonsynonymous nucleotide substitution considering a nucleotide substitution model for 61 sense codons. (Three nonsense codons were eliminated.) Let us consider a pair of sequences of homologous codons and let π_j be the relative frequency of the j-th codon. They assumed that the instantaneous substitution rate (q_{ij}) from codon i to codon j is given by the following equations.

$$q_{ij} = \begin{cases} 0, & \text{if nucleotide change occurs at two or more positions} \\ \pi_j, & \text{for synonymous transversion} \\ k\pi_j, & \text{for synonymous transition} \\ \omega\pi_j, & \text{for nonsynonymous transversion} \\ \omega k\pi_j, & \text{for nonsynonymous transition} \end{cases} \tag{1.34}$$

where k is the transition/transversion rate ratio and ω is the nonsynonymous/synonymous rate ratio. Here k may be written as α/β if the rates of transitional and transversional changes are α and β, respectively.

There are 61 parameters for π_j, but if we assume that the codon frequencies are in equilibrium, they can be estimated by the observed codon frequencies when the number of codons used is large. Therefore, the only parameters to be estimated are k and ω, and these parameters can be estimated by using the maximum likelihood method (Goldman and Yang 1994).

References

Adachi, J., and Hasegawa, M. (1996). Model of amino acid substitution in proteins encoded by mitochondrial DNA. *Journal of Molecular Evolution* 42, 459–468.

Barry, D., and Hartigan, J.A. (1987). Asynchronous distance between homologous DNA sequences. *Biometrics* 43, 261–276.

Bustamante, C.D., Fledel-Alon, A., Williamson, S., Nielsen, R., Hubisz, M.T., Glanowski, S., Tanenbaum, D.M., White, T.J., Sninsky, J.J., Hernandez, R.D., et al. (2005). Natural selection on protein-coding genes in the human genome. *Nature* 437, 1153–1157.

Bustamante, C.D., Townsend, J.P., and Hartl, D.L. (2000). Solvent accessibility and purifying selection within proteins of Escherichia coli and Salmonella enterica. *Molecular Biology and Evolution* 17, 301–308.

Cavender, J.A., and Felsenstein, J. (1987). Invariants of phylogenies in a simple case with discrete states. *Journal of Classification* 4, 57–71.

Chamary, J.V., Parmley, J.L., and Hurst, L.D. (2006). Hearing silence: non-neutral evolution at synonymous sites in mammals. *Nature Reviews Genetics* 7, 98–108.

Dayhoff, M.O. (1972). Atlas of protein sequence and structure 1972. (National Biomedical Research Foundation).

Dayhoff, M.O. (1978). Protein segment dictionary 78. (National Biomedical Research Foundation).

Dean, A.M., and Golding, G.B. (1997). Protein engineering reveals ancient adaptive replacements in isocitrate dehydrogenase. *Proceedings of the National Academy of Sciences of the United States of America* 94, 3104–3109.

Gillespie, J.H. (1991). The causes of molecular evolution (New York, Oxford University Press).

Goldman, N., and Yang, Z. (1994). A codon-based model of nucleotide substitution for protein-coding DNA sequences. *Molecular Biology and Evolution* 11, 725–736.

Gu, X., and Li, W.H. (1992). Higher rates of amino acid substitution in rodents than in humans. *Molecular Phylogenetics and Evolution* 1, 211–214.

Gu, X., and Li, W.H. (1996a). Bias-corrected paralinear and LogDet distances and tests of molecular clocks and phylogenies under nonstationary nucleotide frequencies. *Molecular Biology and Evolution* 13, 1375–1383.

Gu, X., and Li, W.H. (1996b). A general additive distance with time-reversibility and rate variation among nucleotide sites. *Proceedings of the National Academy of Sciences of the United States of America* 93, 4671–4676.

Gu, X., and Li, W.H. (1998). Estimation of evolutionary distances under stationary and nonstationary models of nucleotide substitution. *Proceedings of the National Academy of Sciences of the United States of America* 95, 5899–5905.

Gu, X., Wang, Y., and Gu, J. (2002). Age distribution of human gene families shows significant roles of both large- and small-scale duplications in vertebrate evolution. *Nature Genetics* 31, 205–209.

Hasegawa, M., and Hashimoto, T. (1993). Ribosomal-RNA Trees Misleading. *Nature* 361, 23–23.

Hedges, S.B., and Kumar, S. (2009). The Timetree of Life. (Oxford University Press).

Huang, W., Chang, B.H., Gu, X., Hewett-Emmett, D., Li, W.H. (1997). Sex differences in mutation rate in higher primates estimated from AMG intron sequences. *Journal of Molecular Evolution* 44, 463–465.

Hughes, A.L., and Nei, M. (1988). Pattern of nucleotide substitution at major histocompatibility complex class I loci reveals overdominant selection. *Nature* 335, 167–170.

Jin, L., and Nei, M. (1990). Limitations of the evolutionary parsimony method of phylogenetic analysis. *Molecular Biology and Evolution* 7, 82–102.

Jones, D.T., Taylor, W.R., and Thornton, J.M. (1992). The rapid generation of mutation data matrices from protein sequences. *Computer Applications in the Biosciences* 8, 275–282.

Jukes, T.H., and Cantor, C.R. (1969). Evolution of Protein Molecules. In *Mammalian Protein Metabolism* (Academic Press), 21–132.

Kimura, M. (1968). Evolutionary rate at the molecular level. *Nature* 217, 624–626.

Kimura, M. (1980). A simple method for estimating evolutionary rates of base substitutions through comparative studies of nucleotide sequences. *Journal of Molecular Evolution* 16, 111–120.

Kimura, M. (1983). Rare variant alleles in the light of the neutral theory. *Molecular Biology and Evolution* 1, 84–93.

Kimura, M., and Ohta, T. (1971). Protein polymorphism as a phase of molecular evolution. *Nature* 229, 467–469.

Kumar, S., and Hedges, S.B. (1998). A molecular timescale for vertebrate evolution. *Nature* 392, 917–920.

Lake, J.A. (1994). Reconstructing evolutionary trees from DNA and protein sequences: paralinear distances. *Proceedings of the National Academy of Sciences* 91, 1455–1459.

Lanave, C., Preparata, G., Saccone, C., and Serio, G. (1984). A new method for calculating evolutionary substitution rates. *Journal of Molecular Evolution* 20, 86–93.

Lee, Y.H., Ota, T., and Vacquier, V.D. (1995). Positive Selection Is a General Phenomenon in the Evolution of Abalone Sperm Lysin. *Molecular Biology and Evolution* 12, 231–238.

Li, W.H. (1993). Unbiased estimation of the rates of synonymous and nonsynonymous substitution. *Journal of Molecular Evolution* 36, 96–99.

Li, W.H., Gu, X. (1996). Estimating evolutionary distances. In *Methods in Enzymology* (Academic Press) 266, 449–459.

Li, W.H. (1997). Molecular evolution (Sunderland, Mass., Sinauer Associates).

Li, W.H., Wu, C.I., and Luo, C.C. (1985). A new method for estimating synonymous and non-synonymous rates of nucleotide substitution considering the relative likelihood of nucleotide and codon changes. *Molecular Biology and Evolution* 2, 150–174.

Lockhart, P.J., Steel, M.A., Hendy, M., and Penny, D. (1994). Recovering evolutionary trees under a more realistic model of sequence evolution. *Molecular Biology and Evolution* 11, 605–612.

McDonald, J.H., and Kreitman, M. (1991). Adaptive protein evolution at the Adh locus in Drosophila. *Nature* 351, 652–654.

Messier, W., Stewart, C.B. (1997). Episodic adaptive evolution of primate lysozymes. *Nature*, 385, 151–154.

Miyata, T., and Yasunaga, T. (1980). Molecular evolution of mRNA: a method for estimating evolutionary rates of synonymous and amino acid substitutions from homologous nucleotide sequences and its application. *Journal of Molecular Evolution* 16, 23–36.

Nei, M., and Gojobori, T. (1986) Simple methods for estimating the numbers of synonymous and nonsynonymous nucleotide substitutions. *Molecular Biology and Evolution*, 3, 418–426.

Nei, M. (1987). *Molecular Evolutionary Genetics* (New York: Columbia University Press).

Rodríguez, F., Oliver, J.L., Marín, A., and Medina, J.R. (1990). The general stochastic model of nucleotide substitution. *Journal of Theoretical Biology* 142, 485–501.

Nei, M. and Kumar, S. (2000). Molecular Evolution and Phylogenetics. (Oxford University Press).

Nielsen, R., Hellmann, I., Hubisz, M., Bustamante, C., and Clark, A.G. (2007). Recent and ongoing selection in the human genome. *Nature Reviews Genetics* 8, 857–868.

Ohta, T. (1973) Slightly deleterious mutant substitutions in evolution. *Nature* 246, 96–98.

Ohta, T. (1993). An examination of the generation-time effect on molecular evolution. *Proceedings of the National Academy of Sciences of the United States of America* 90, 10676–10680.

Perler, F., Efstratiadis, A., Lomedico, P., Gilbert, W., Kolodner, R., and Dodgson, J. (1980). The Evolution of Genes - the Chicken Preproinsulin Gene. *Cell* 20, 555–566.

Saitou, N., and Nei, M. (1987). The neighbor-joining method: a new method for reconstructing phylogenetic trees. *Molecular Biology and Evolution* 4, 406–425.

Sogin, M.L., Hinkle, G., and Leipe, D.D. (1993). Universal Tree of Life. *Nature* 362, 795–795.

Steel, M.A. (1994). Recovering a tree from the leaf colourations it generates under a Markov model. *Applied Mathematics Letters* 7, 19–23.

Tajima, F., and Nei, M. (1982). Biases of the estimates of DNA divergence obtained by the restriction enzyme technique. *Journal of Molecular Evolution* 18, 115–120.

Tajima, F., and Nei, M. (1984). Estimation of evolutionary distance between nucleotide sequences. *Molecular Biology and Evolution* 1, 269–285.

Tamura, K., and Nei, M. (1993). Estimation of the number of nucleotide substitutions in the control region of mitochondrial DNA in humans and chimpanzees. *Molecular Biology and Evolution* 10, 512–526.

Tanenbaum, D.M., Civello, D., White, T.J., et al. (2005). A scan for positively selected genes in the genomes of humans and chimpanzees. *PLoS Biology* 3, e170.

Wang, Y., and Gu, X. (2001). Functional divergence in the caspase gene family and altered functional constraints: statistical analysis and prediction. *Genetics* 158, 1311–1320.

Wu, C.I., and Li, W.H. (1985). Evidence for higher rates of nucleotide substitution in rodents than in man. *Proceedings of the National Academy of Sciences of the United States of America* 82, 1741–1745.

From Atlas of Protein Sequence and Structure.

Zhang, J., Rosenberg, H.F., and Nei, M. (1998). Positive Darwinian selection after gene duplication in primate ribonuclease genes. *Proceedings of the National Academy of Sciences of the United States of America* 95, 3708–3713.

Zharkikh, A. (1994). Estimation of evolutionary distances between nucleotide sequences. *Journal of Molecular Evolution* 39, 315–329.

2
Phylogeny-Dependent Analysis

2.1 Phylogenetics trees: an overview

Phylogenetic relationships of genes or organisms (generally referred to as taxa) are usually presented in a tree-like form, either with rooted or without any root. These can be referred to as rooted trees or unrooted trees. The branching pattern of a tree is called the topology. If the number of taxa (m) is four, there are 15 possible rooted tree topologies and 3 possible unrooted tree topologies. However, the number of possible topologies rapidly increases with increasing m (millions of possible trees when $m \gg 15$). Therefore, it becomes a difficult task to find the true tree topology when m is large; the subject of study is called phylogenetic inference.

Use of molecular data to reconstruct phylogenetic trees can be traced back to Cavalli-Sforza and Edwards (1967) and Fitch and Margoliash (1967). Since a DNA sequence splits into two descendant sequences at the time of speciation or gene duplication, molecular phylogenetic trees are usually bifurcating. In an unrooted bifurcating tree of m taxa there are $2m - 3$ branches. Since there are m exterior branches connecting to m extent taxa, the number of interior branches is $m - 3$. The number of interior nodes is equal to $m - 2$. In a rooted tree, the numbers of interior branches and interior nodes are $m - 2$ and $m - 1$, respectively, and the total number of branches is $2m - 2$. However, when a relatively short sequence is considered, some interior branches may show no nucleotide substitution, so a multifurcating node may appear. This type of tree is called a multifurcating tree.

In phylogenetic inference, an optimization principle, such as the maximum likelihood or the minimum evolution principle, is often used for choosing the most likely topology. There are many methods that have been developed for reconstructing phylogenetic trees from molecular data, which can be classified into four major groups: (1) distance methods, (2) parsimony methods, (3) likelihood methods, and (4) Bayesian methods. When the phylogeny is inferred, we need to evaluate the statistical reliability. For the

Statistical Analysis of Molecular and Genomic Evolution. Xun Gu, Oxford University Press. © Xun Gu (2024).
DOI: 10.1093/oso/9780198816515.003.0002

distance, parsimony, or likelihood methods, the bootstrapping approach (Felsenstein 1985) has been widely-used. In the following sections we discuss each of them briefly. Due to space limitations, we will not address some advanced topics, such as phylogenetic inference under the nonhomogeneous model of DNA sequence evolution (Galtier and Gouy 1995, 1998) or maximum likelihood identification of coevolving protein residues (Pollock et al. 1999).

2.2 Distance method for phylogenetic inference

2.2.1 Principle: minimum-evolution

In distance methods, evolutionary distances are computed for all pairs of taxa, and a phylogenetic tree is constructed by considering the relationships among these distance values. There are many different methods of constructing trees from distance data. Here we discuss a method that has been widely used in molecular evolution: the principle of minimum-evolution (ME) and the neighbor-joining (NJ) algorithm. In this method, the sum (S) of all branch length estimates in a given topology, that is,

$$S = \sum_{i=1}^{T} \hat{b}_i \qquad (2.1)$$

where $\hat{b}_i$ is the estimated branch length of the i-th branch, is computed for all plausible topologies, and the topology that has the smallest S value is chosen as the best tree. The theoretical foundation of the ME method is Rzhetsky and Nei's (1993) mathematical proof that when unbiased estimates of evolutionary distances are used, the expected value of S becomes the smallest for the true topology.

2.2.2 Algorithm: neighbor joining (NJ) method

Although the ME method has satisfactory statistical properties, it requires a substantial amount of computer time when the number of taxa compared is large. Saitou and Nei (1987) developed an efficient tree-building method that is based on the minimum evolution principle. This method does not examine all possible topologies, but at each stage of taxon clustering a minimum evolution principle is used. This method is called the neighbor joining (NJ) method. It should be noticed that there are a number of versions that may improve the performance of NJ algorithm, for example BIONJ (Gascuel 1997).

Initial condition of NJ algorithm: star-tree

Construction of a tree by the NJ method begins with a star tree, which is produced under the assumption that there is no clustering of taxa (star-tree) (Fig. 2.1). Then we estimate the branch lengths of the star tree and compute the sum of all branches (S_0). Mathematically, S_0 for the star-tree is given by

$$S_0 = \sum_{i=1}^{m} L_{iX} = \sum_{i<j}^{m} d_{ij}/(m-1) = T/(m-1) \tag{2.2}$$

Where L_{iX} is the branch length estimate between nodes i and X, and $T = \sum_{i<j} d_{ij}$. In the star-tree, i stands for the i-th exterior node and X the interior node. Since the star-tree is generally incorrect, this sum (S_0) should be greater than the sum (S_F) for the final NJ tree.

Identification of the first neighbor pair

Since we do not know which pair of taxa are true neighbors, we have to consider each pair of taxa as a potential pair of neighbors and compute the sum of branch lengths (S_{ij}) for the i-th and j-th taxa using a topology similar to that given in Fig. 2.1. Suppose, as shown in Fig. 2.1, that taxa 1 and 2 are neighbors, such that S_{12} is given by the sum of the following terms

$$S_{12} = L_{1X} + L_{2X} + L_{XY} + \sum_{i=3}^{m} L_{iY} \tag{2.3}$$

where $L_{1X} + L_{2X} = d_{12}$. Saitou and Nei (1987) showed that S_{12} can be calculated as follows

$$S_{12} = \frac{1}{2(m-2)} \sum_{i=3}^{m} (d_{1i} + d_{2i}) + \frac{1}{2} d_{12} + \frac{1}{m-2} \sum_{3 \le i < j} d_{ij} \tag{2.4}$$

Obviously, S_{ij} can be computed in the same way if we replace 1 and 2 by i and j, respectively. We then choose taxa i and j that show the smallest S_{ij} value.

Reduction of star-tree

Once a pair of neighbors with the smallest S_{ij} is identified, the two are combined into one composite taxon. Then we can create a new node (A) that connects taxa i and j. The branch lengths (b_{A_i} and b_{A_j}) from this node to taxon i and j are given by

$$b_{Ai} = \frac{1}{2(m-2)} [(m-2)d_{ij} + R_i - R_j]$$

$$b_{Aj} = \frac{1}{2(m-2)} [(m-2)d_{ij} - R_i + R_j] \tag{2.5}$$

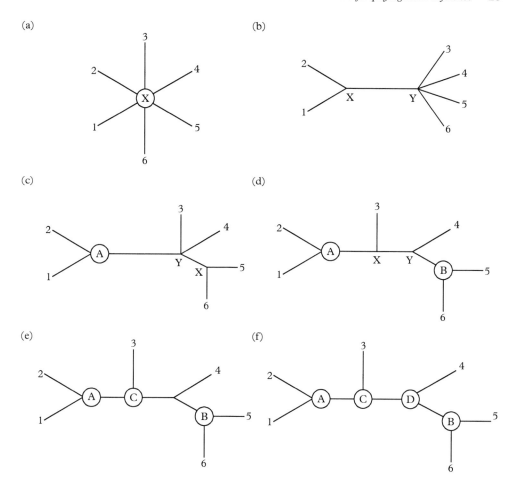

Figure 2.1 *Illustration of the computational procedure in the neighbor-joining (NJ) method. The NJ method begins with a start tree (a). Once a pair of neighbors (taxa 1 and 2) is identified (b), they are combined into one composite taxon called taxon A (c). For the rest of four original taxa plus composite taxon A, suppose taxa 5 and 6 are the neighbor so that they are again combined to be composite taxon B (d). The final two steps (e and f) solves the tree-making problem in a similar iteration.*

where $R_i = \sum_{k=1}^{m} d_{ik}$ and $R_j = \sum_{k=1}^{m} d_{jk}$. These values are known to be least squares estimates for the topology under consideration (Saitou and Nei 1987).

Iteration

The next step is to compute the distance between the new node (A) and the remaining taxa $(k \neq i,j)$ (Fig. 2.1). This distance is given by

$$d_{Ak} = (d_{ik} + d_{jk} - d_{ij})/2 \qquad (2.6)$$

If we compute all the distances using this equation, we have a new $(m-1) \times (m-1)$ matrix. From this matrix, we can compute a new S_{ij} matrix. To find the new pair of "neighbors," we choose a pair with the smallest S'_{ij} value. A new node B is then created for this pair of taxa, and a new $(m-2) \times (m-2)$ distance matrix is computed. This procedure is repeated until all taxa are clustered in a single unrooted tree. The final tree obtained in this way is the NJ tree.

2.2.3   Four-point conditions and the NJ algorithm

Fitch's criteria

Consider a tree for four taxa and assume that taxa 1 and 2 are a pair of true neighbors. For an additive tree, we obviously have the following inequalities

$$d_{12} + d_{34} \quad < \quad d_{13} + d_{24}$$
$$d_{12} + d_{34} \quad < \quad d_{14} + d_{23} \tag{2.7}$$

This four-point condition has been used by Fitch (1981) to reconstruct the topology of a tree. It should be noted that the NJ method and these two methods require the same condition for obtaining the correct topology for the case of four taxa. To show this, we compare the difference between S_{13} (the sum of branch lengths after assuming taxa 1 and 3 are the pair of neighbors) and S_{12} (the sum after assuming taxa 1 and 2 are the pair of neighbors). It has been shown that

$$S_{13} - S_{12} = \frac{1}{2(m-2)} \sum_{k=4}^{m} [(d_{13} + d_{2k}) - (d_{12} + d_{3k})] = \frac{1}{2(m-2)} \sum_{k=4}^{m} U_{12,3k} \tag{2.8}$$

where $U_{12,3k} = d_{13} + d_{2k} - (d_{12} + d_{3k})$ is the score of four-point condition. On the right hand of this equation is the sum of four-point conditions between taxa 1, 2, 3, and any of $3 < k \leq m$. Therefore, the NJ criterion for choosing taxa 1 and 2 as a pair of neighbors, that is, $S_{13} > S_{12}$, is the same condition that the sum of four-point condition scores is positive.

The Q-score of Studier and Keppler

Studier and Keppler (1988) rewrote the formula of S_{12} as follows

$$S_{12} = \frac{2T - R_1 - R_2}{2(m-2)} + \frac{d_{12}}{2} \tag{2.9}$$

where $R_i = \sum_{k=1}^{m} d_{ik}$ and $R_j = \sum_{k=1}^{m} d_{jk}$. Since T is the same for all pairs, S_{12} can be replaced by

$$Q_{12} = (m-2)d_{12} - R_i - R_j \qquad (2.10)$$

for the purpose of computing the relative value of S_{ij} (Studier and Keppler 1988). In fact, most computer programs use Q_{12} rather than S_{12} to facilitate the computation.

2.3 Parsimony methods for phylogenetic inference

2.3.1 Terminologies

Parsimony

Maximum parsimony (MP) methods were originally developed for morphological characters (Hennig 1966). Here we discuss how MP can be useful for analyzing molecular data. These MP methods consider four or more aligned nucleotide (or amino acid) sequences ($m \geq 4$). The smallest number of nucleotide (or amino acid) substitutions that explain the entire evolutionary process for the topology is computed. This computation is done for all potential topologies, and the topology that requires the smallest number of substitutions, also called the shortest tree length, is chosen as the best tree. The theoretical basis of this method is William of Ockham's philosophical idea that the best hypothesis to explain a process is the one that requires the smallest number of assumptions.

Homoplasy and long-branch attraction

If there are no backward and no parallel substitutions (no homoplasy) at any nucleotide/amino acid site, MP methods are expected to produce the correct tree. However, nucleotide/amino acid sequences are often subject to backward and parallel substitutions (high homoplasy). When this occurs, MP methods tend to give incorrect topologies. Felsenstein (1978) has shown that when the rate of nucleotide substitution varies extensively among branches, MP methods may generate incorrect topologies. When this occurs, long branches of the true tree tend to join (or attract) together in the MP inferred tree. Therefore, this phenomenon is called the long-branch attraction.

Unweighted MP and weighted MP methods

In unweighted MP methods, all types of nucleotide or amino acid substitutions are assumed to occur with nearly-equal probability. However, certain substitutions such as transitions occur more often than transversion substitutions. It is therefore reasonable to assign different weights to different types of substitutions, giving rise to the so-called weighted MP methods.

Informative sites

In MP methods, nucleotide or amino acid sites that have the same type of nucleotide or amino acid sites (invariable sites) are eliminated from the analysis. It should be noted that not all variable sites are useful. For instance, any nucleotide sites at which only unique nucleotides exist (called singleton sites) can be explained by the same substitution pattern in all topologies. For a nucleotide site to be informative for constructing an MP tree, there must be at least two different kinds of nucleotides, each represented in at least two sequences (taxa). These sites are called informative sites, or more precisely, parsimony-informative sites. In the MP tree reconstruction, it is sufficient to consider only parsimony-informative sites.

Consistency index and homoplasy index

It is important to have many informative sites to obtain reliable MP trees. However, when the extent of homoplasy (backward and parallel substitutions) is high, MP trees would not be reliable because these informative sites tend to be inconsistent. For this reason, several measures have been proposed to calculate the extent of homoplasy. In systematics, a widely used measure is the consistency index (CI) computed for all informative sites, and the homoplasy index $HI = 1 - CI$. When there are no backward or parallel substitutions, we have $CI = 1$ and $HI = 0$. In this case, the topology is uniquely determined by the principle of parsimony.

2.3.2   Searching for MP trees

When the number of sequences or taxa (m) is small, say, $m < 10$, it is possible to compute the tree lengths of all possible trees and determine the MP tree. This type of search for MP trees is called an exhaustive search. While the number of topologies rapidly increases as m, it is virtually impossible to examine all topologies if m is large. However, if we know clearly incorrect topologies, we can simply compute only for potentially correct trees. This type of search is called a specific-tree search.

There are two ways of obtaining MP trees when $m > 10$. One is to use the branch-and-bound method (Hendy and Penny 1982). In this method, the trees that obviously have a tree length longer than that of a previously examined tree are all ignored, and the MP tree is determined by evaluating the tree lengths for a group of trees that potentially have shorter tree lengths. This method guarantees finding all MP trees, although it is not an exhaustive search. However, even this method becomes very time-consuming if m is large. In this case, one has to use another approach called the heuristic search. In this method, only a small portion of all possible trees is examined, and there is no guarantee that the MP tree will be found. However, it is possible to enhance the probability of obtaining the MP tree by using several heuristic algorithms.

2.4 Maximum likelihood (ML) methods for phylogenetic inference

Felsenstein (1981) has established the basic framework to infer the phylogenetic tree based on the principle of maximum likelihood. Later, Yang (1993, 1994a, 1994b, 1997)has extended this approach to tackle a number of issues in molecular evolution. In the following we use a simple example to illustrate the maximum likelihood (ML) methods for phylogenetic inference.

2.4.1 Likelihood function

Consider a simple tee of four taxa (DNA sequences with n nucleotides and no deletions/insertions) given in Fig. 2.2. Denote the observed nucleotides (A, T, C, or G) for sequences 1, 2, 3, and 4 at a given site by x_1, x_2, x_3, and x_4, respectively. The unobserved nucleotides at root or internal nodes 0, 5, and 6 are denoted by x_0, x_5, and x_6, respectively.

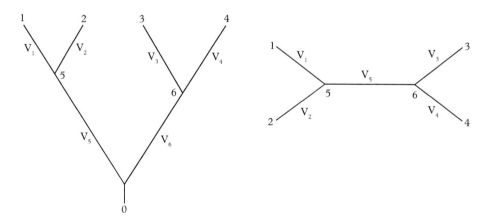

Figure 2.2 *A hypothetical phylogenetic tree (rooted and unrooted, respectively) has been used to illustrate the building of likelihood function in the text. In both cases, two internal nodes are numbered by 5 and 6, while four external branch lengths are denoted by v_1, v_2, v_3 and v_4, respectively. In the rooted tree, there are two internal branch lengths toward to the ancestor O, with branch lengths v_5 and v_6, respectively, while these two branches are merged into one in the unrooted tree.*

Let $P_{ij}(t)$ be the probability that nucleotide i at the initial time becomes nucleotide j at time t at a given site. Here i and j refer to any of A, T, C, and G. In the ML method, the rate of substitution (r) is allowed to vary from branch to branch, so that it is convenient to measure evolutionary time in terms of the expected number of substitutions ($v = rt$). In the following, we denote the expected number of substitutions for the i-th branch by $v_i = r_i t_i$. According to the Markov-chain property, the likelihood function for a nucleotide site (k-th) is then given by

$$l_k = \pi_{x_0} P_{x_0 x_5}(v_5) P_{x_0 x_6}(v_6) P_{x_5 x_1}(v_1) P_{x_5 x_2}(v_2) P_{x_6 x_3}(v_3) P_{x_6 x_4}(v_4)$$

where π_{x_0} is the probability that node 0 (root) has nucleotide x_0, which is often set to be the same as the relative frequency of nucleotide x_0 in the entire set of sequences. In practice we do not know x_0, x_5, and x_6, so the likelihood will be the sum of the above quantity overall possible nucleotides at root 0 and internal nodes 5 and 6. That is,

$$L_k = \sum_{x_0} \sum_{x_5} \sum_{x_6} \pi_{x_0} P_{x_0 x_5}(v_5) P_{x_0 x_6}(v_6) P_{x_5 x_1}(v_1) P_{x_5 x_2}(v_2) P_{x_6 x_3}(v_3) P_{x_6 x_4}(v_4) \qquad (2.11)$$

In ML methods, one must consider all nucleotide sites including invariable sites. Since the likelihood (L) for the entire sequence is the product of L_k's for all sites, the log likelihood of the entire tree becomes

$$\ln L = \sum_{k=1}^{n} \ln L_k$$

2.4.2   Time-reversibility and the root problem

To know $P_{ij}(v)$ explicitly, we have to use a specific substitution model. For instance, Felsenstein (1981) used the equal-input model, in which $P_{ii}(v)$ and $P_{ij}(v)$ ($i \neq j$) are given by

$$P_{ii}(v) = \pi_i + (1 - \pi_i)e^{-v}$$
$$P_{ij}(v) = \pi_j(1 - e^{-v})$$

where π_i is the relative frequency of the i-th nucleotide. When $\pi_i = 1/4$ and $v = 4rt$, the above equations become identical with those for the Jukes–Cantor model.

In particular, we consider a class of substitution models called time-reversible. This is because if we use a reversible model of nucleotide substitution for defining $P_{ij}(v)$, there is no need to consider the root (Fig. 2.2). A reversible model means that the process of nucleotide substitutions between time 0 and time t remains the same whether we consider the evolutionary process forward or backward in time. Mathematically, the reversibility condition is given by $\pi_i P_{ij}(v) = \pi_j P_{ji}(v)$ for all i and j. One can easily verify that the equal-input model satisfies this condition. When a reversible model is used, the number of nucleotide substitutions $(v_5 + v_6)$ between nodes 5 and 6 of tree A remains the same irrespective of the location of root 0. Therefore, we designate $v_5 + v_6$ in tree A by v_5 in tree B. Assuming that evolutionary change starts from some point of the tree, say, from node 5 for convenience, the likelihood function L_k can be rewritten in the following way,

$$L_k = \sum_{x_5} \sum_{x_6} \pi_{x_5} P_{x_5 x_6}(v_5) P_{x_5 x_1}(v_1) P_{x_5 x_2}(v_2) P_{x_6 x_3}(v_3) P_{x_6 x_4}(v_4) \qquad (2.12)$$

2.4.3 Search strategies for ML trees

In practice, we have to consider all nucleotide sites. Since the likelihood (L) for the entire sequence is the product of all sites, the log likelihood for a topology in general may be written as

$$\ln L = \sum_{k=1}^{n} \ln L_k = f(\mathbf{x}, \theta)$$

where $\mathbf{x}$ is a set of observed nucleotide sequences and θ is a set of parameters, such as branch lengths, nucleotide frequencies, and substitution rates. In ML methods, the likelihood of observing a given set of sequence data for a specific substitution model is maximized for each topology, and the topology that gives the highest maximum likelihood is chosen as the final tree.

Since the search for an ML tree is time-consuming, various heuristic methods for finding the ML tree have been proposed (e.g., Felsenstein 1981; Adachi and Hasegawa 1996). Though many of these algorithms, in principle, are similar to those used for obtaining minimum evolution or parsimony trees, their efficiencies in obtaining the correct topology are not necessarily the same.

For distantly-related protein-coding genes, the DNA likelihood may have some problems, because the synonymous substitution may have been saturated, suggesting that the stationary model of nucleotide substitution is no longer valid. In this case, the evolutionary change of protein sequences may be more appropriate. Kishino et al. (1990) proposed a protein-likelihood method, which used Dayhoff (1978) empirical transition matrix for 20 different amino acids. Later, Adachi and Hasegawa (1996) used various transition matrices including the Poisson model, Jones et al.'s (1992) empirical transition matrix for nuclear proteins, and their own matrix for mitochondrial proteins.

2.5 Bayesian methods for phylogenetic inference

Bayesian methods provide a computationally efficient approach to infer the phylogenetic tree, by calculating the posterior distribution of phylogenetic trees (Huelsenbeck et al. 2001). Given the multiple alignment of sequence data, $\mathbf{D}$, according to the Bayes rule, it has been stated that the posterior probability of the i-th possible tree topology denoted by T_i is given by

$$P(T_i|\mathbf{D}) = \frac{P(T_i)P(\mathbf{D}|T_i)}{\sum_T P(T)P(\mathbf{D}|T)} \tag{2.13}$$

where $P(T_i|\mathbf{D})$ is the probability of tree T_i given the sequence data $\mathbf{D}$, $P(\mathbf{D}|T_i)$ is the probability or likelihood of the data given tree T_i, and $P(T_i)$ is the prior probability of T_i. The denominator sums the probabilities over all possible trees. Moreover, for each of these possible trees, $P(\mathbf{D}|T_i)$ has to be integrated over all possible values of the branch lengths of the tree and over the parameters of the model of sequence evolution. Let $\mathbf{t}$ be a vector of the branch lengths of the tree and $\mathbf{m}$ a vector of the parameters of the model of sequence evolution, then we have

$$P(\mathbf{D}|T_i) = \int_{\mathbf{t}} \int_{\mathbf{m}} P(\mathbf{D}|T_i, \mathbf{t}, \mathbf{m})P(\mathbf{t})P(\mathbf{m})d\mathbf{t}d\mathbf{m} \tag{2.14}$$

where $P(\mathbf{t})$ and $P(\mathbf{m})$ are the prior probabilities of the branch lengths and the parameters of the model.

It should be noted that the number of possible unrooted topologies for n species is $(2n-5)!/2^{n-3}(n-3)!$. That means the summation in the denominator is over a high number of topologies even for ten sequences. The problem can be solved via the MCMC sampling algorithm (Hastings 1970; Metropolis et al. 1953). Indeed, under the MCMC methods, a Markov-chain is constructed, the states of which are different phylogenetic trees (Huelsenbeck et al. 2001; Larget and Simon 1999; Mau et al. 1999; Rannala and Yang 1996; Yang and Rannala 1997). At each step in the chain, a new tree is proposed by altering the topology, or by changing branch lengths, or by changing the parameters of the model of sequence evolution. The Metropolis–Hastings algorithm is then used to accept or reject the new tree. A newly proposed tree that improves upon the previous tree in the chain is always accepted; otherwise it is accepted with probability proportional to the ratio of its likelihood to that of the previous tree in the chain. If such a Markov-chain has been run long enough, it reaches a stationary distribution.

The argument for Bayesian inference of phylogeny can be concisely stated as follows. At stationarity, the Metropolis–Hastings sampling algorithm ensures that the Markov chain wanders through the universe of trees, sampling better and worse trees, rather than inexorably moving toward "better" trees as an optimizing approach would do. A properly constructed chain samples trees from the posterior density of trees in proportion to their frequency of occurrence in the actual density. That is, the Markov chain draws a sample of trees that can be used to approximate the posterior distribution. In the current implementation, the stationary distribution simultaneously samples the posterior density of trees, the posterior distribution of the branch lengths, and parameters of the model of sequence evolution. With desirable degree of precision, in practice the chain should be allowed to run perhaps hundreds of thousands or million of trees.

2.6   Ancestral sequence inference

Ancestral sequence inference has been shown useful to predict the ancestral function of a gene family, to detect positive selection, and to reconstruct the ancestral genome (Golding and Dean 1998; Soyer et al. 2006). There are two major types: the parsimony approach and the probabilistic approach, which are briefly discussed below.

2.6.1   The maximum parsimony approach

The idea of maximum parsimony is to identify the ancestral states at each node of a tree that minimize the number of character (nucleotide or amino acid) changes needed to explain the observed differences among the extant sequences. We use the original algorithm developed by Fitch (1971) as an example (for nucleotides) to show the principle. To exemplify the Fitch algorithm, we consider a simple five-taxon tree in Fig. 2.3. For the character illustrated, the data observed are $X_1 = A$, $X_2 = C$, $X_3 = G$, $X_4 = C$, and $X_5 = T$. At node X_6, the intersection of the sets of its two descendants X_1 and X_2 is $(A) \cap (C) = \phi$ (here ϕ means empty). If it is empty, the union set must be assigned X_6: $(A) \cup (C) = (A, C)$. Likewise, the union set of X_3 and X_4 is assigned at node X_7 with $(G) \cup (C) = (G, C)$. Now the set at node X_8 can be determined, since the intersection of sets X_5 and X_7 is again empty, the union of these sets (G, C, T) is assigned. Finally the set in the root X_9 is the intersection of the sets X_8 and X_6: $(A, C) \cap (G, C, T) = (C)$. Three union operations were needed, suggesting a minimum of three changes is needed for this reconstruction.

In the next step, the ancestral states are determined (Fig. 2.3) by traversing the tree in pre-order (from the root to the extant taxon, or leaves). First the state C is determined at the root, the state at X_8 is also set to C because this state is in the ancestral state in the parent node (X_9) and also is a member of set at that node (X_8). Similarly, the state at X_6 is C and the state at X_7 is also C. Because the Fitch algorithm penalizes

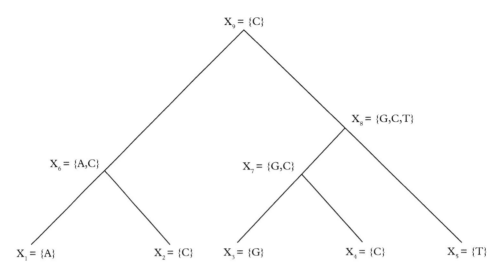

Figure 2.3 *Given the current observed nucleotide pattern in five sequences at a site (denoted by X_1 to X_5), the ancestral state of each internal node (X_6 to X_9) can be inferred by the parsimony algorithm or the probabilistic method.*

equally any change among the four nucleotide states, the procedure can result in multiple ancestral state inferences with equal parsimony. The Sankoff (1975) algorithm is a generalization of Fitch's original version, allowing different costs for different character changes.

2.6.2   The probabilistic (Bayesian) approach

Though the parsimony method is simple and intuitive, the major shortcoming is the lack of statistical evaluation for the reconstructed ancestral states. For instance, the parsimony method is not statistically meaningful for discriminating among equally parsimonious reconstructions. To solve this problem, a number authors (e.g., Schluter 1995; Yang et al. 1995; Koshi and Goldstein 1996; Pupko et al. 2000, 2002) developed efficient algorithms, using the probabilistic models.

The probabilistic (Bayesian) approach for ancestral state inference has two steps in principle. We use Fig. 2.3 for illustration. First, it calculates the likelihood (the joint probability) of the observed nucleotide pattern $X_1, \ldots, X_5 = (A, C, G, C, T)$ under the given phylogeny. Denote the four nucleotides by 1, 2, 3, and 4, respectively, it formally can be written as

$$P(X_1, \ldots, X_5) = \sum_{x_9=1}^{4} f_{x_9} \sum_{x_8=1}^{4} \sum_{x_7=1}^{4} \sum_{x_6=1}^{4} P_{x_9 x_6} P_{x_9 x_8} P_{x_6 x_1} P_{x_6 x_2} P_{x_8 x_5} P_{x_8 x_7} P_{x_7 x_4} P_{x_7 x_5}$$

where f_{x_9} is the nucleotide frequency at the root (X_9); $P_{x_i x_j}$ is the transition probability from nodes X_i to X_j; here branch lengths are omitted for simplicity. Note that in most cases, the probabilistic model for ancestral sequence inference is time-reversible. As a result, the position of a root does not affect the likelihood so that the ancestral root state cannot be reconstructed. Rewrite $P(X_1, \ldots, X_5)$ as

$$P(X_1, \ldots, X_5) = \sum_{x_8=1}^{4} \sum_{x_7=1}^{4} \sum_{x_6=1}^{4} P(X_1, \ldots, X_5; x_6, x_7, x_8)$$

where $P(X_1, \ldots, X_5; x_6, x_7, x_8) = \sum_{x_9=1}^{4} f_{x_9} P_{x_9 x_6} P_{x_9 x_8} P_{x_6 x_1} P_{x_6 x_2} P_{x_8 x_5} P_{x_8 x_7} P_{x_7 x_4} P_{x_7 x_5}$. That is, the likelihood $P(X_1, \ldots, X_5)$ is a sum of $4 \times 4 \times 4 = 64$ different terms, each corresponding to a specific ancestral sequence assignment at nodes X_6, X_7, and X_8. Therefore, the joint nucleotide assignment $(x_6, x_7,$ and $x_8)$ which contributes the most to the above likelihood is called the ancestral state inference, which is explicitly given by the expression

$$\text{argmax}_{x_6, x_7, x_8} \left[P(X_1, \ldots, X_5; x_6, x_7, x_8) \right]$$

The maximum of the above expression is the likelihood of the joint reconstruction. In essence, the posterior probability of x_6, x_7, and x_8 given the dataset $(X_1, \ldots, X_5)$ is

$$P(x_6, x_7, x_8 | X_1, \ldots, X_5) = \frac{P(X_1, \ldots, X_5; x_6, x_7, x_8)}{P(X_1, \ldots, X_5)}$$

In other words, one may also view the joint nucleotide assignment $(x_6, x_7,$ and $x_8)$ which has the highest posterior probability given the data as the ancestral state inference.

In practice, particularly when the number of sequence is large, the computational time required to determine the joint likelihood (or posterior probability) becomes huge. One feasible approach is to use single-node inference. This algorithm considers the nucleotide assignment for each internal node separately. For instance, one may write the likelihood with respect to node X_8

$$P(X_1, \ldots, X_5) = \sum_{x_8=1}^{4} \left[\sum_{x_7=1}^{4} \sum_{x_6=1}^{4} P(X_1, \ldots, X_5; x_6, x_7, x_8) \right]$$

Hence, the nucleotide assignment at node X_8 which contributes the most to the likelihood is given by

$$\text{argmax}_{x_8} \left[\sum_{x_7=1}^{4} \sum_{x_6=1}^{4} P(X_1, \ldots, X_5; x_6, x_7, x_8) \right]$$

It has been showed the assignment of characters (nucleotides or amino acids) at ancestral nodes could be affected by the substitution model of nucleotides or amino acids. For instance, two almost equal likelihood assignments under the simple Jukes–Cantor (1969) model may have different outcomes under more sophisticated models. Nevertheless, ancestral sequence inference is fairly robust against other parameter estimation such as branch lengths or the rate variation among sites, though the statistical evaluation for the inference reliability could vary; see Yang et al. (1995) and Pupko et al. (2002) for detailed discussions.

2.6.3   Deletions and insertions

None of the models described thus far consider deletions and insertions (called indels or gaps). A common technique in phylogenetic inference is to exclude all sites in which at least one sequence contains a gap. However, for ancestral sequence inference, the goal is to infer the most likely, nearly complete ancestral sequence. To this end, it is essential to determine whether a character (nucleotide or amino acid) or gap is the ancestral state. This problem has not been solved effectively. In the following we introduce some tentative approaches.

Gap as an additional character: One approximation to escape this problem is to represent a gap by adding an additional character to the model (thus creating an alphabet of 21 characters for amino acids, or five for DNA/RNA). In spite of easy implementation, there are two main difficulties with this approach: First, the probabilities of such transitions from each character state to a gap and *vice versa* are unclear. Second, this approach assumes independence among sites. Thus, an insertion or deletion at two sites will be considered as two independent so-called character-to-gap transitions, rather than the biologically more reasonable explanation of a single two-site indel event.

Edward-Shields method: The algorithm developed by Edwards-Shields (2004) first approximates the probabilities of gaps at each site and internal node, using a two-state character model (0 is a gap site and 1 for a non-gap site). Once the ancestral state (0/1) for each node was determined, the nongapped sites are estimated in an informal likelihood approach using probabilities derived from empirical substitution matrices.

An integrated framework: Recently, we proposed an integrated approach to infer a nearly complete ancestral sequence (unpublished results). The novelty of this approach is in dividing the ancestral sequence inference algorithm into several separate tasks as follows:

(1) For nucleotide or amino acid sites without any gap, follow the previous probabilistic method to assign the most likely character states at internal nodes.

(2) Given the phylogeny, count the number distribution of indels with the gap length. Assume that the change of indels follow a Poisson process with the parameter $v_k T$ (v_k is the rate of indels with gap length k, and T is the total evolutionary time of the tree). According to the study of Gu and Li (1995), we use the log-link function $v_k = v_0/(1 + b\ln k)$ ($k \geq 1$) to measure the length-dependent evolutionary rate of indels. Apparently, v_k is inversely related to the gap length. The unknown parameters, v_0 and b, can be estimated from the size distribution of indels.

(3) For a given site with indels, we adopt the additional character approach with an important extension. That is, the rate from other characters to the gap or *vice versa* is proportional to the gap-length factor $v_0/(1 + b\ln k)$.

We have conducted some preliminary analysis and found that the performance is, in general, satisfactory, as long as the multiple alignment is reliable.

2.7   Rate variation among sites

It is well known that different amino acid residues of a protein may have different functional constraints such that the substitution rate varies among sites. Although this phenomenon was first described over 50 years ago (Uzzell and Corbin 1971), its importance for molecular evolutionary study has not been recognized until recently (Nei and Kumar 2000; Yang 2007; Gu 1999, 2001, 2007a, 2007b). In particular, the gamma distribution has been widely used for modeling the rate variation among sites (Yang 1993; Gu and Li 1995). Under this model, the variation of substitution rate (λ) among sites can be described as follows:

$$\phi(\lambda) = \frac{\beta^\alpha}{\Gamma(\alpha)}\lambda^{\alpha-1}e^{-\beta\lambda}$$

where the shape parameter α is important because it describes the degree of rate variation, and β is a scalar. Since $1/\sqrt{\alpha}$ is the coefficient of variation of λ, the larger α is, the weaker the rate variation is, and $\alpha = \infty$ means a uniform rate among sites.

Several methods have been developed for estimating α from sequence data, these methods can be classified into two groups. The first group use the maximum likelihood (ML) approach, which was constructed under the framework of Felsenstein (1981) (e.g., Yang 1993; Gu and Li 1995). However, the algorithms developed by these authors for maximizing the likelihood function are time-consuming. This problem has been solved by the approximate method of discrete-gamma distribution (Yang 1994b). The second group of methods for estimating α rely on the parsimony method, which has been widely used because it is computationally fast (e.g., Uzzell and Corbin 1971, Holmquist et al. 1983; Tamura and Nei 1993; Sullivan et al. 1995; Tourasse and Gouy

1997). In these methods, the principle of parsimony (Fitch 1971) was used to infer the (minimum-required) number of substitutions. Since the parsimony method tends to underestimate the number of substitutions, it is known that the shape parameter (α) can be seriously overestimated: in other words, the degree of rate variaton among sites can be underestimated (Wakeley 1993).

Since rate variation among sites has been shown important for evolutionary functional analysis of protein families (Gu 1999, 2001, 2006), as well as the estimation of gene pleiotropy (Gu 2007a), implementation of a statistically unbiased and computationally fast method is desirable. Gu and Zhang (1997) proposed a simple ML method that has two steps: (1) At each site, the expected number of substitutions corrected for multiple hits is estimated by a likelihood approach, based on the phylogeny and inferred ancestral sequences; and (2) the ML estimate of α is obtained under a negative binomial distribution (Uzzell and Corbin 1971) using the estimated number of substitutions.

2.7.1   Number of substitutions at a site

The number of substitutions at a site cannot be observed from the present-day sequences, so it has to be inferred. The traditional method of inference invokes the parsimony principle (Fitch 1971), which tends to underestimate the true number of substitutions (Gu and Zhang 1997; Zhang and Gu 1998). To understand the bias caused by parsimony, it is important to distinguish between the number of substitutions (k) and the number of branches on which the amino acids (or nucleotides) at the two ends of a branch are different (m); in the following, m is also concisely called the number of changes. For given sequence data with a known tree, the difference between m and k is due to multiple substitutions that may occur when the branch is long, resulting in $m \leq k$. It should be noted that the minimum-required number of substitutions is actually an inference of m rather than k, because the possibility of multiple hits is completely neglected.

Gu and Zhang (1997) developed an asymptotically unbiased method to estimate k. For simplicity, we only discuss amino acid sequences here; it is virtually the same for nucleotide sequences. Suppose we have a protein dataset with n homologous sequences, whose phylogenetic tree (topology) is known or can be inferred. It is known that the total number of branches for an unrooted tree is $M = 2n - 3$, or $M = 2n - 2$ for a rooted tree. At a given site, we assume that k along the tree follows a Poisson distribution, whose expectation is written by $\bar{k} = uB$, where B is the total branch length of the tree and u is the evolutionary rate at this site. It is noteworthy that k under the Poisson model is a random variable. Our purpose is to estimate the expected number of substitutions ($\bar{k}$), which will be used for estimating α.

The number of substitutions at a given site that occur on branch i also follows a Poisson distribution with the expectation ub_i, where u is the site-specific rate and b_i is the length of branch i. Because the expected number of substitutions at this site along the phylogeny is $\bar{k} = uB$, we have $ub_i = \bar{k}b_i/B$. Thus, the probability of no change on branch i (i.e., the amino acids at the two ends of this branch are the same) is given by

$$p_i = \exp\{-\bar{k}b_i/B\}$$

and the probability of a change (i.e., the amino acids are different at the two ends of the branch) is

$$q_i = 1 - p_i = 1 - \exp\{-\bar{k}b_i/B\}$$

For a given site, the branches along the tree can be divided into two groups. The first group, denoted by G_1, includes the branches on which (amino acid) changes occur, and the second group, denoted by G_0, includes the branches on which no changes occur. Obviously, the total number of branches in G_1 is equal to m at the site, and the total number of branches on G_0 is therefore given by $M - m$. Then, when the information about groups G_1 and G_0 at a site is known, the (conditional) likelihood function can be written as

$$L = \prod_{i \in G_1} q_i \prod_{j \in G_0} p_j = \prod_{i \in G_1} \left[1 - \exp\{-\bar{k}b_i/B\}\right] \prod_{j \in G_0} \exp\{-\bar{k}b_j/B\} \qquad (2.15)$$

The subscripts under the product signs mean that branch i belongs to group G_1 and branch j belongs to group G_0, respectively.

After re-numbering branches in group G_1 from 1 to m and branches in group G_0 from $m+1$ to M, Gu and Zhang (1997) has shown that the ML estimate equation with respect to $\bar{k}$ can be concisely expressed by

$$\sum_{i=1}^{m} \frac{b_i/B}{1 - e^{-\hat{k}b_i/B}} = 1 \qquad (2.16)$$

The ML estimate of the expected number of substitutions ($\hat{k}$), which is the positive solution of the above equation, depends on m and the estimated branch lengths. If for every branch, b_i/B is so small that $1 - e^{-\hat{k}b_i/B} \approx \hat{k}b_i/B$, then $\hat{k} \approx m$. This result is consistent with the intuition that the expected number of substitutions estimated approaches the number of changes when all branch lengths in the tree are short. On the other hand, if all branch lengths are the same, i.e., $b_i = b$ and so $B = Mb$ (M is the number of branches), it can be simplified as follows

$$\hat{k} = -M\ln\left(1 - \frac{m}{M}\right) \tag{2.17}$$

In the above formulation, we assume that ancestral sequences are known. Thus, under a given phylogenetic tree, it is easy to classify a branch into G_0 or G_1 by simply comparing the amino acids at the two ends of the branch at each site, and count the number of changes along the tree (m). In practice, ancestral sequences have to be inferred by the Bayesian-based methods (e.g., Schluter 1995; Yang et al. 1995; Koshi and Goldstein 1996; Zhang et al. 1997). In this approach, amino acid assignment that has the highest (posterior) probability is chosen to represent the inferred ancestral amino acids at this site.

2.7.2   Estimation of α

If amino acid (or nucleotide) substitutions at each site follow a Poisson process, and the substitution rate (λ) varies among sites according to the gamma distribution $\phi(\lambda)$, the number of sites with the occurrence of k substitutions follows a negative binomial distribution, i.e.,

$$f(k) = \frac{\Gamma(\alpha + k)}{k!\Gamma(\alpha)}\left(\frac{D}{D+\alpha}\right)^k\left(\frac{\alpha}{D+\alpha}\right)^\alpha \tag{2.18}$$

where D is the average number of substitutions per site along the tree.

The ML approach for estimating the parameter α from a negative binomial distribution was clearly discussed by Johnson and Kotz (1969); it was also used by Sullivan et al. (1995) and Tourasse and Gouy (1997) for the rate variation among sites. In our case, the difference from the standard algorithm is that the number of substitutions at a site is replaced by its expectation $\hat{k}$. Therefore, the log-likelihood function can be written as

$$\ln L = \sum_{i=1}^{N} \ln f(\hat{k}_i)$$

where N is the total number of sites and $\hat{k}$ is the estimate of the expected number of substitutions at site i, which is not necessarily an integer. One can easily show that the ML estimate of D is the same as that for the normal case, which is given by $\hat{D} = \sum_{i=1}^{N} \hat{k}_i/N$. There is no simple solution for the estimate of α, but it can be numerically obtained, and the sampling variance of α can also be approximately obtained.

References

Adachi, J., and Hasegawa, M. (1996). Model of amino acid substitution in proteins encoded by mitochondrial DNA. *Journal of Molecular Evolution* 42, 459–468.

Cavalli-Sforza, L.L., and Edwards, A.W. (1967). Phylogenetic analysis. Models and estimation procedures. *American Journal of Human Genetics* 19, 233–257.

Dayhoff, M.O. (1978). Protein segment dictionary 78. (National Biomedical Research Foundation).

Edwards, R.J., and Shields, D.C. (2004). GASP: Gapped Ancestral Sequence Prediction for proteins. *BMC Bioinformatics* 5, 123.

Felsenstein, J. (1978). Cases in which Parsimony or Compatibility Methods Will be Positively Misleading. *Systematic Biology* 27, 401–410.

Felsenstein, J. (1981). Evolutionary trees from DNA sequences: a maximum likelihood approach. *Journal of Molecular Evolution* 17, 368–376.

Felsenstein, J. (1985). Confidence limits on phylogenies: an approach using the bootstrap. *Evolution* 39, 783–791.

Fitch, W.M., and Margoliash, E. (1967). Construction of phylogenetic trees. *Science* 155, 279–284.

Fitch, W.M. (1971). Toward defining the course of evolution: minimum change for a specific tree topology. *Systematic Biology* 20, 406–416.

Fitch, W.M. (1981). A non-sequential method for constructing trees and hierarchical classifications. *Journal of Molecular Evolution* 18, 30–37.

Galtier, N., and Gouy, M. (1995). Inferring phylogenies from DNA sequences of unequal base compositions. *Proceedings of the National Academy of Sciences of the United States of America* 92, 11317–11321.

Galtier, N., and Gouy, M. (1998). Inferring pattern and process: maximum-likelihood implementation of a nonhomogeneous model of DNA sequence evolution for phylogenetic analysis. *Molecular Biology and Evolution* 15, 871–879.

Gascuel, O. (1997). BIONJ: an improved version of the NJ algorithm based on a simple model of sequence data. *Molecular Biology and Evolution* 14, 685–695.

Golding, G.B., and Dean, A.M. (1998). The structural basis of molecular adaptation. *Molecular Biology and Evolution* 15, 355–369.

Gu, X. (1999). Statistical methods for testing functional divergence after gene duplication. *Molecular Biology and Evolution* 16, 1664–1674.

Gu, X. (2001). Maximum-likelihood approach for gene family evolution under functional divergence. *Molecular Biology and Evolution* 18, 453–464.

Gu, X. (2006). A simple statistical method for estimating type-II (cluster-specific) functional divergence of protein sequences. *Molecular Biology and Evolution* 23, 1937–1945.

Gu, X. (2007a). Evolutionary framework for protein sequence evolution and gene pleiotropy. *Genetics* 175, 1813–1822.

Gu, X. (2007b). Stabilizing selection of protein function and distribution of selection coefficient among sites. *Genetica* 130, 93–97.

Gu, X., and Li, W.H. (1995). The size distribution of insertions and deletions in human and rodent pseudogenes suggests the logarithmic gap penalty for sequence alignment. *Journal of Molecular Evolution* 40, 464–473.

Gu, X., and Zhang, J. (1997). A simple method for estimating the parameter of substitution rate variation among sites. *Molecular Biology and Evolution* 14, 1106–1113.

Gu, X., Fu, Y.-X., and Li, W.-H. (1995). Maximum likelihood estimation of the heterogeneity of substitution rate among nucleotide sites. *Molecular Biology and Evolution* 12, 546–557.

Hastings, W.K. (1970). Monte-Carlo sampling methods using Markov chains and their applications. *Biometrika* 57, 97–109.

Hendy, M.D., and Penny, D. (1982). Branch and bound algorithms to determine minimal evolutionary trees. *Mathematical Biosciences* 59, 277–290.

Hennig, W. (1966). Phylogenetic Systematics. (University of Illinois Press).

Holmquist, R., Goodman, M., Conroy, T., and Czelusniak, J. (1983). The spatial distribution of fixed mutations within genes coding for proteins. *Journal of Molecular Evolution* 19, 437–448.

Huelsenbeck, J.P., Ronquist, F., Nielsen, R., and Bollback, J.P. (2001). Bayesian inference of phylogeny and its impact on evolutionary biology. *Science* 294, 2310–2314.

Johnson, N.L., and Kotz, S. (1969). Discrete distributions (Bostonorr, Houghton Mifflin).

Jones, D.T., Taylor, W.R., and Thornton, J.M. (1992). The rapid generation of mutation data matrices from protein sequences. *Computer Applications in the Biosciences* 8, 275–282.

Jukes, T.H., and Cantor, C.R. (1969). Evolution of Protein Molecules. In *Mammalian Protein Metabolism*, (Academic Press), 21–132.

Kishino, H., Miyata, T., and Hasegawa, M. (1990). Maximum likelihood inference of protein phylogeny and the origin of chloroplasts. *Journal of Molecular Evolution* 31, 151–160.

Koshi, J.M., and Goldstein, R.A. (1996). Probabilistic reconstruction of ancestral protein sequences. *Journal of Molecular Evolution* 42, 313–320.

Larget, B., and Simon, D.L. (1999). Markov chain Monte Carlo algorithms for the Bayesian analysis of phylogenetic trees. *Molecular Biology and Evolution* 16, 750–759.

Mau, B., Newton, M.A., and Larget, B. (1999). Bayesian phylogenetic inference via Markov chain Monte Carlo methods. *Biometrics* 55, 1–12.

Metropolis, N., Rosenbluth, A.W., Rosenbluth, M.N., Teller, A.H., and Teller, E. (1953). Equation of state calculations by fast computing machines. *Journal of Chemical Physics* 21, 1087–1092.

Nei, M. and Kumar, S. (2000). Molecular Evolution and Phylogenetics. (Oxford University Press).

Pollock, D.D., Taylor, W.R., and Goldman, N. (1999). Coevolving protein residues: maximum likelihood identification and relationship to structure. *Journal of Molecular Biology* 287, 187–198.

Pupko, T., Pe'er, I., Hasegawa, M., Graur, D., and Friedman, N. (2002). A branch-and-bound algorithm for the inference of ancestral amino-acid sequences when the replacement rate varies among sites: Application to the evolution of five gene families. *Bioinformatics* 18, 1116–1123.

Pupko, T., Pe'er, I., Shamir, R., and Graur, D. (2000). A fast algorithm for joint reconstruction of ancestral amino acid sequences. *Molecular Biology and Evolution* 17, 890–896.

Rannala, B., and Yang, Z. (1996). Probability distribution of molecular evolutionary trees: a new method of phylogenetic inference. *Journal of Molecular Evolution* 43, 304–311.

Rzhetsky, A., and Nei, M. (1993). Theoretical foundation of the minimum-evolution method of phylogenetic inference. *Molecular Biology and Evolution* 10, 1073–1095.

Saitou, N., and Nei, M. (1987). The neighbor-joining method: a new method for reconstructing phylogenetic trees. *Molecular Biology and Evolution* 4, 406–425.

Sankoff, D. (1975). Minimal mutation trees of sequences. *SIAM Journal on Applied Mathematics* 28, 35–42.

Schluter, D. (1995). Uncertainty in ancient phylogenies. *Nature* 377, 108–109.

Soyer, O.S., and Bonhoeffer, S. (2006). Evolution of complexity in signaling pathways. *Proceedings of the National Academy of Sciences of the United States of America* 103, 16337–16342.

Studier, J.A., and Keppler, K.J. (1988). A note on the neighbor-joining algorithm of Saitou and Nei. *Molecular Biology and Evolution* 5, 729–731.

Sullivan, J., Holsinger, K.E., and Simon, C. (1995). Among-site rate variation and phylogenetic analysis of 12S rRNA in sigmodontine rodents. *Molecular Biology and Evolution* 12, 988–1001.

Tamura, K., and Nei, M. (1993). Estimation of the number of nucleotide substitutions in the control region of mitochondrial DNA in humans and chimpanzees. *Molecular Biology and Evolution* 10, 512–526.

Tourasse, N.J., and Gouy, M. (1997). Evolutionary distances between nucleotide sequences based on the distribution of substitution rates among sites as estimated by parsimony. *Molecular Biology and Evolution* 14, 287–298.

Uzzell, T., and Corbin, K.W. (1971). Fitting discrete probability distributions to evolutionary events. *Science* 172, 1089–1096.

Wakeley, J. (1993). Substitution rate variation among sites in hypervariable region 1 of human mitochondrial DNA. *Journal of Molecular Evolution* 37, 613–623.

Yang, Z. (1993). Maximum-likelihood estimation of phylogeny from DNA sequences when substitution rates differ over sites. *Molecular Biology and Evolution* 10, 1396–1401.

Yang, Z. (1994a). Estimating the pattern of nucleotide substitution. *Journal of Molecular Evolution* 39, 105–111.

Yang, Z. (1994b). Maximum likelihood phylogenetic estimation from DNA sequences with variable rates over sites: approximate methods. *Journal of Molecular Evolution* 39, 306–314.

Yang, Z. (1997). PAML: a program package for phylogenetic analysis by maximum likelihood. *Computer Applications in the Biosciences: CABIOS* 13, 555–556.

Yang, Z., Kumar, S., and Nei, M. (1995). A new method of inference of ancestral nucleotide and amino acid sequences. *Genetics* 141, 1641–1650.

Yang, Z. (2007). PAML 4: phylogenetic analysis by maximum likelihood. *Molecular Biology and Evolution* 24, 1586—1591.

Yang, Z., and Rannala, B. (1997). Bayesian phylogenetic inference using DNA sequences: a Markov Chain Monte Carlo Method. *Molecular Biology and Evolution* 14, 717–724.

Yang, Z., Kumar, S., and Nei, M. (1995). A new method of inference of ancestral nucleotide and amino acid sequences. *Genetics* 141, 1641–1650.

Zhang, J., and Gu, X. (1998). Correlation between the substitution rate and rate variation among sites in protein evolution. *Genetics* 149, 1615–1625.

Zhang, J., Kumar, S., and Nei, M. (1997). Small-sample tests of episodic adaptive evolution: a case study of primate lysozymes. *Molecular Biology and Evolution* 14, 1335–1338.

3

Neutrality Testing when Some Amino Acid Sites are Strongly Constrained

The ratio of the substitution rate of nonsynonymous sites to that of synonymous sites, denoted by d_N/d_S, has provided a simple way of assessing the selection pressure on protein coding sequences: the ratio less than 1 reflecting negative selection and that greater than 1 reflecting positive selection. Investigators have developed a rich body of methods for estimating d_N/d_S ratios [see Nielsen (2005) for a review]. However, the interpretation of d_N/d_S ratio analysis becomes complicated when different types of selection are imposed on amino acid sites simultaneously. In this chapter, we discuss some statistical methods that enables us to distinguish between positive selection, neutrality, and nearly-neutral evolution when some amino acid sites are subject to strong functional constraints.

3.1 The $\triangle$-criterion on protein sequence evolution

3.1.1 Evolutionary rate of an encoding-gene

In the theory of molecular evolution, the evolutionary rate (λ) of a nucleotide is determined by three levels of genetic processes: mutation at the individual level, polymorphism at the population level, and fixation at the species level (Kimura 1962, 1983). Let v be the mutation rate, s the coefficient of selection, and N_e the effective population size. It is well-known that the evolutionary rate can be written as follows

$$\lambda = v\frac{4N_e s}{1 - e^{-4N_e s}} \tag{3.1}$$

It appears that the selection intensity $S = 4N_e s$ uniquely determines the ratio of evolutionary rate to the mutation rate λ/v. Eq. (3.1) provided a theoretical platform for the

Statistical Analysis of Molecular and Genomic Evolution. Xun Gu, Oxford University Press. © Xun Gu (2024).
DOI: 10.1093/oso/9780198816515.003.0003

debate between adaptive evolution, neutral evolution, and nearly-neutral evolution: it predicts $\lambda/v > 1$ for adaptive evolution ($s > 0$), $\lambda/v = 1$ for neutral evolution ($s = 0$), or $\lambda/v < 1$ for deleterious evolution ($s < 0$), respectively.

In the study of protein sequence evolution, the most relevant to Eq. (3.1) is the ratio of nonsynonymous distance (d_N) to synonymous distance (d_S) (Li et al. 1985; Nei and Gojobori 1986; Li 1993; Goldman and Yang 1994; Muse and Gaut 1994; Yang 1997). Under the assumption that synonymous substitutions in the coding region are selectively neutral, the d_N/d_S ratio has been widely used as an empirical proxy to the ratio of evolutionary rate to the mutation rate: $d_N/d_S = 1$ means a strict neutral evolution in nonsynonymous substitutions; $d_N/d_S > 1$ for positive selection and $d_N/d_S < 1$ for negative selection (Nielsen 2005). However, the null hypothesis of $d_N/d_S = 1$ (strictly neutral) can be confounded by some amino acid sites subject to very strong purifying selections (Ohta 1992). Therefore, the observation of $d_N/d_S < 1$ (for most genes in most organisms) can be explained by either strict neutrality, nearly-neutrality, or even positive selection, with the existence of some invariable amino acid sites.

Gu (2022) tackled this issue. Since each nonsynonymous site may have different selection intensity ($S = 4N_e s$), it was generally assumed that S varies among nonsynonymous sites according to a distribution $\Phi(S)$, and a constant mutation rate (v). Hence, the d_N/d_S ratio is expected to be the mean rate of nonsynonymous substitution (scaled by the mutation rate v), that is,

$$\frac{d_N}{d_S} \sim \frac{E[\lambda]}{v} = \int \left[\frac{S}{1 - e^{-S}}\right] \Phi(S) dS \tag{3.2}$$

where $E[.]$ is the mean evolutionary rate.

3.1.2 Derivation of the Δ-criterion

It has been realized that Eq. (3.2) is not sufficient to distinguish different selection scenarios when some amino acid sites are invariable. Instead, Gu (2022) considered the second moment of the evolutionary rate, that is,

$$\frac{E[\lambda^2]}{v^2} = \int \left[\frac{S}{1 - e^{-S}}\right]^2 \Phi(S) dS \tag{3.3}$$

A new quantity Δ is then defined as the difference between the second-moment and the mean of evolutionary rate, scaled by the mutation rate. From Eq. (3.2) and Eq. (3.3), it is given by

$$\Delta = \frac{E[\lambda^2]}{v^2} - \frac{E[\lambda]}{v} = \int \left(\frac{S}{1 - e^{-S}}\right) \left(\frac{S}{1 - e^{-S}} - 1\right) \Phi(S) dS \tag{3.4}$$

As shown below, the sign of Δ provides insights on different selection scenarios.

$\Delta = 0$ *under the neutral-lethal selection mode*

Under the classical theory of neutrality (Kimura 1983), $\Phi(S)$ can be constructed as a binary distribution: all mutations are classified into two categories, strictly neutral mutations ($S = 0$ with a proportion of $1 - f_L$) or lethal mutations ($S = -\infty$ with a proportion of f_L). Since the evolutionary rate is $\lambda = v$ (mutation rate) for strictly neutral mutations and $\lambda = 0$ for lethal mutations, the first (mean) and the second moments of evolutionary rate of a gene under this model are simply as follows

$$E[\lambda] = (1 - f_L)v$$
$$E[\lambda^2] = (1 - f_L)v^2 \qquad (3.5)$$

In short, under the neutral-lethal selection scenario, we have $\Delta = 0$, regardless of the proportion of invariable sites.

$\Delta > 0$ *under positive selection with neutral-lethal mode*

Under this mode, all mutations of a gene are classified into three categories: adaptive mutations ($S > 0$ with a probability of f_A), lethal mutations ($S = -\infty$ with a probability of f_L), and neutral mutations ($S = 0$ with a probability of $1 - f_A - f_L$). Let $\Phi^+(S)$ be the distribution of (positive) selection intensity for adaptive mutations. It follows that the mean and the second moments of the evolutionary rate are given by

$$E[\lambda] = f_A v \int_0^\infty \left[\frac{S}{1 - e^{-S}}\right] \Phi^+(S)dS + (1 - f_A - f_L)\, v$$
$$E[\lambda^2] = f_A v^2 \int_0^\infty \left[\frac{S}{1 - e^{-S}}\right]^2 \Phi^+(S)dS + (1 - f_A - f_L)\, v^2 \qquad (3.6)$$

respectively. One can further verify that the Δ-measure given by Eq. (3.4) can be written as follows

$$\Delta = f_A \int_0^\infty \left(\frac{S}{1 - e^{-S}}\right)\left(\frac{S}{1 - e^{-S}} - 1\right) \Phi^+(S)dS > 0 \qquad (3.7)$$

which is always larger than 0, regardless of the existence of lethal or neutral mutations.

$\Delta < 0$ *under nearly-neutral mode with lethal mutations*

Under this mode, all mutations of a gene are classified into three categories: nearly-neutral mutations ($S < 0$ with a probability of $1 - f_L - f_0$), lethal mutations ($S = -\infty$ with a probability of f_L), and neutral mutations ($S = 0$ with a probability of f_0).

Let $\Phi^-(S)$ be the distribution of (negative) selection intensity of those nearly-neutral mutations. Similar to Eqs.(3.6) and (3.7), we have

$$E[\lambda] = (1 - f_0 - f_L)v \int_{-\infty}^{0} \left[\frac{S}{1 - e^{-S}}\right] \Phi^-(S)dS + f_0 v$$

$$E[\lambda^2] = (1 - f_0 - f_L)v^2 \int_{-\infty}^{0} \left[\frac{S}{1 - e^{-S}}\right]^2 \Phi^-(S)dS + f_0 v^2 \qquad (3.8)$$

respectively, and therefore

$$\Delta = (1 - f_0 - f_L) \int_{-\infty}^{0} \left(\frac{S}{1 - e^{-S}}\right) \left(\frac{S}{1 - e^{-S}} - 1\right) \Phi^-(S)dS < 0 \qquad (3.9)$$

which is always less than 0.

3.2 The d_N/d_S-H ratio test based on the Δ-criterion

3.2.1 Theoretical formulation

The d_N/d_S-H test developed by Gu (2022) invoked the H-measure defined by

$$H = 1 - \frac{(E[\lambda])^2}{E[\lambda^2]} = 1 - \frac{(E[\lambda]/v)^2}{E[\lambda^2]/v^2} \qquad (3.10)$$

Ranging from 0 to 1, a high value of H indicates a high degree of rate variation among amino acid sites, and *vice versa*. After the replacement of $E[\lambda]/v$ by d_N/d_S in Eq. (3.10), we have

$$\frac{E[\lambda^2]}{v^2} = \frac{(E[\lambda]/v)^2}{1 - H} = \frac{(d_N/d_S)^2}{1 - H} \qquad (3.11)$$

It follows that the Δ-measure can be written in terms of H and d_N/d_S, that is

$$\Delta = \frac{d_N}{d_S} \left(\frac{d_N/d_S}{1 - H} - 1\right) \qquad (3.12)$$

Eq. (3.12) reveals a simple connection between Δ and d_N/d_S: $\Delta < 0$ indicates $d_N/d_S < 1 - H$, $\Delta = 0$ indicates $d_N/d_S = 1 - H$, and $\Delta > 0$ indicates $d_N/d_S > 1 - H$, respectively. Accordingly, one can formulate the d_N/d_S–H test as follows (Table 3.1).

(*i*) The null hypothesis is $d_N/d_S = 1 - H$, as expected by the strictly neutral evolution while some sites are subject to very strong purifying selections (invariable sites).

Table 3.1 *A summary of d_N/d_S–H analysis*

Criteria	Interpretations
d_N/d_S <1-H	Nearly-neutral evolution, plus some sites under strong selective constraints
d_N/d_S =1-H	Neutral evolution, plus some sites under strong selective constraints
1> d_N/d_S >1-H	Positive evolution, plus some sites under strong selective constraints
d_N/d_S =1	Neutral evolution virtually in all sites (or a combination between positive and negative selections)
d_N/d_S >1	Dominant positive evolution

    (*ii*) Rejection of the null hypothesis because $d_N/d_S > 1 - H$ suggests a positive selection whereas some functionally important sites are virtually invariable.

    (*iii*) Rejection of the null hypothesis because $d_N/d_S < 1 - H$ provides the statistical evidence for the nearly-neutral model instead of the strictly-neutral model with some invariable sites.

In short, the d_N/d_S-H test has two goals: (*i*) to detect the signal of positive selection in an encoding gene with strong functional constraints; and (*ii*) to distinguish between the nearly neutral model and the strictly-neutral model with invariable sites.

3.2.2 Implementation

Similar to most d_N/d_S tests, the d_N/d_S-H analysis focuses on two pre-specified coding sequences between closely related species or duplicates, referred as *the target species pair*. The main issue here is to estimate H, which requires the phylogenetic analysis of the same orthologous genes from more distantly-related species. Let $E[x]$ and $Var(x)$ be the mean and variance of the number of changes per amino acid site, respectively. Suppose that amino acid changes at a site follow a Poisson process, whereas the evolutionary rate (λ) varies among sites. Then one may write $E[x] = E[\lambda]T$ and $V(x) = V(\lambda)T^2 + E[\lambda]T$, respectively, where $V(\lambda) = E[\lambda^2] - (E[\lambda])^2$ is the variance of evolutionary rate among sites, and T is the total evolutionary time along the phylogeny. By equating $E[x]$ and $V(x)$ with their sampling mean $\bar{x}$ and the sampling variance $Var(x)$, respectively, one can show a simple estimate of H by

$$\hat{H} = \frac{Var(x) - \bar{x}}{Var(x) + \bar{x}(\bar{x} - 1)} \tag{3.13}$$

The procedure of d_N/d_S-H test can be briefly described as follows.

(*i*) Estimate the d_N/d_S ratio by any published method from a given pair of closely-related species (target species pair), such as Li et al. (1985), Nei and Gojobori (1986), Li (1993), and Yang (1997). One can calculate the sampling variance $Var(d_N/d_S)$ based on the original paper.

(*ii*) Consider an inferred phylogeny of a multiple alignment of protein sequences that also includes the target species pair. The parsimony method is then used to infer the minimum-required number (*m*) of changes per site. This number is further corrected by the method of Gu and Zhang (1997) to reduce the under-estimation: if all branch lengths of the phylogeny are identical, for instance, the corrected number (*x*) of substitutions at a site is given by obtained by

$$x = -M \ln(1 - m/M) \tag{3.14}$$

where M is the number of branches of the phylogeny.

(*iii*) After the sampling mean and variance of x over all sites are calculated, it is straightforward to estimate H by Eq. (3.13).

(*iv*) Note that the large sampling variance of the coefficient of variation CV is approximately by

$$Var(CV) \approx \frac{CV^2(1 + 2CV^2)}{2L} \tag{3.15}$$

where L is the number of amino acid sites. The sampling variance of H is approximated by the delta-method.

(*v*) Let $Z = d_N/d_S - (1 - H)$. Under the assumption that d_N/d_S and H are independent, the large sampling variance of Z is given by $Var(Z) = Var(d_N/d_S) + Var(H)$. Then a simple Z-test is applied to test whether the null hypothesis

$$d_N/d_S = (1 - H) \tag{3.16}$$

can be statistically rejected.

3.2.3   Case studies

Gu (2022) analyzed three independent datasets: (*a*) 342 vertebrate genes from eight species (the target species pair is the human-mouse); (*b*) 437 Drosophila genes from 12 species (the target species pair is *D. melanogaster-D.yakuba*); and (c) 580 yeast genes from five species (the target species pair is *S. cerevisiae-C. glabrata*). In each dataset, the majority of genes satisfy the condition of $d_N/d_S < 1 - H$ with p-value< 0.05, suggesting that nearly-neutral evolution (fixation of slightly deleterious mutations) dominated the protein evolution in vertebrates, flies, and yeasts (Fig. 3.1). Meanwhile, a minor

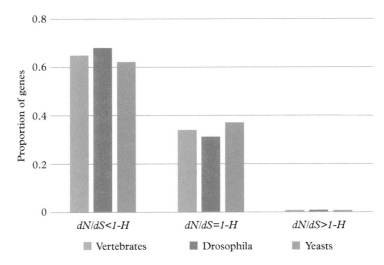

Figure 3.1 *The d_N/d_S-H analysis for three datasets: 342 vertebrate genes from eight species (the target species pair is human-mouse); 437 Drosophila genes from 12 species (the target species pair is D. melanogaster-D.yakuba); and 580 yeast genes from five species (the target species pair is S. cerevisiae-C. glabrata). In each dataset, a gene belongs to the category of d_N/d_S=1-H is determined by the p-value>0.05 under the null hypothesis; otherwise to the category of $d_N/d_S < 1$-H or $d_N/d_S > 1$-H accordingly.*

Material from: Gu, dN/dS-H, a New Test to Distinguish Different Selection Modes in Protein Evolution and Cancer Evolution, Journal of Molecular Evolution, published 2022, Springer Nature

but nontrivial portion of genes showed that the null hypothesis $d_N/d_S = 1 - H$ cannot be statistically rejected at the significance level of 0.05. Hence, strict neutral evolution with strong negative selection may explain the evolutionary pattern in a small portion of proteins. These findings were further confirmed by a larger data set that includes 4336 vertebrate genes, each of which has one-to-one orthology among human mouse, dog, cow, chicken, Xenopus, fugu, and zebrafish. Strikingly, almost all of genes demonstrated $d_N/d_S < 1 - H$ (Fig. 3.2). We thus conclude that the nearly-neutral model is sufficient to explain the genome-wide pattern of protein evolution in vertebrates.

3.3   Model evaluation and extensions

3.3.1   Power analysis of the d_N/d_S-H test

Gu (2022) carried out extensive simulations to evaluate the statistical performance of the d_N/d_S-H test. The main results are summarized as follows.

(*i*) In the simulation theme with no positive selection, the d_N/d_S-H test can efficiently distinguish between nearly-neutral mutations and lethal mutations

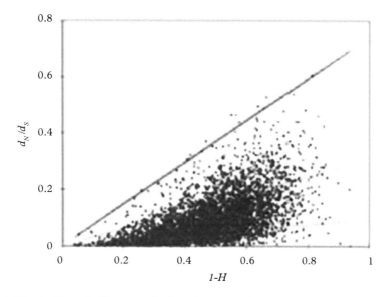

Figure 3.2 *The $d_N/d_S \sim (1\text{-}H)$ scatter plotting of 4336 one-by-one vertebrate genes, where d_N/d_S of each gene is estimated between the human and mouse, and H is estimated from the vertebrate phylogeny. It shows that almost all of genes are below the line of $d_N/d_S = 1\text{-}H$, or $d_N/d_S < 1\text{-}H$. Two lines for 75% and 25% quantiles of the ratio $[d_N/d_S]/[1\text{-}H]$ are also presented.*

Material from: Gu, dN/dS-H, a New Test to Distinguish Different Selection Modes in Protein Evolution and Cancer Evolution, Journal of Molecular Evolution, published 2022, Springer Nature

as long as the sequence length is reasonably long (more than 300 amino acid sites). On the other hand, the power curve shows that the original d_N/d_S test is consistently more powerful than the $d_N/d_S\text{-}H$ test. This observation is expected since the original d_N/d_S-test detects the overall negative effects of lethal and nearly-neutral mutations. Indeed, rejection of the null $d_N/d_S = 1$ is more powerful than that of $d_N/d_S = 1 - H$.

(*ii*) In the simulation theme with positive selection, the $d_N/d_S\text{-}H$ test can effectively detect a weak positive selection under the backgrounds of neutral and lethal mutations. Meanwhile, the power of original d_N/d_S-test could be compromised by the existence of lethal mutations. Nevertheless, there is little power difference between the $d_N/d_S\text{-}H$ test and the d_N/d_S test in the case of strong positive selection.

(*iii*) When the nearly neutral evolution is nontrivial, the power of $d_N/d_S\text{-}H$ test to detect positive selection could be compromised. It appears that the power of $d_N/d_S\text{-}H$ test to detect a weak positive selection is decreased by the existence of nearly-neutral mutations, whereas the original d_N/d_S test is decreased by both weak and strong deleterious mutations.

3.3.2   Relatedness to the branch-site model and tertiary windowing approach

The d_N/d_S-H test can be extended by combining with other methods. A class of methods, termed "branch-site" tests (Yang 1997), offered a model-based phylogenetic hypothesis testing framework for deciding whether or not a lineage (or lineages) of interest had undergone any adaptive change. Branch-site tests measure selective pressure by ω, the ratio of nonsynonymous to synonymous substitution rates, and if a proportion of sites in the sequence provides statistically significant support for $\omega > 1$ along the lineages of interest, then episodic positive selection is inferred. One may extend the d_N/d_S-H test under the phylogenetic framework of "branch-site" tests by replacing ω with $\omega/(1 - H)$: it is theoretically plausible yet may be technically difficult.

Using the leptin gene family as an example, Berglund et al. (2005) proposed a simple method for the examination of d_N/d_S ratios based upon windows in tertiary structure. This tertiary sequence windowing detects new sites under positive diversifying selection and detects positive diversifying selection with a more significant signal along various branches. A simple replacement of d_N/d_S by $d_N/d_S/(1 - H)$ is practically feasible and can make some biologically meaningful identifications after taking the site-specific factor of functional importance of sites into account. To avoid a large sampling effect, one may estimate H estimated from the whole protein sequences.

3.3.3   Extension to cancer somatic mutations: the C_N/C_S-H test

The theory of clonal evolution in cancer biology claims that cancer cells emerge through random somatic mutations from a single cell and genetically diverge to disparate cell subclones and successive clones in cancer cell replication (Stratton et al. 2009; Vogelstein et al. 2013). Ultimately, cells alter one or few crucial pathways and acquire the hallmarks of cancer through somatic mutations followed by cancer-specific positive selections (Ding et al. 2018). The argument that carcinogenesis is a form of evolution at the level of somatic cells suggests that our understanding of cancer initiation and progression can benefit from molecular evolutionary approaches (Bailey et al. 2021). One well-known case is to estimate the rate ratio of somatic nonsynonymous to synonymous somatic mutations (C_N/C_S) of a protein-encoding gene in cancers, but the results were inconsistent (Dees et al. 2012; Lawrence et al. 2013; Reimand et al. 2013; Porta-Pardo and Godzik 2014; Mularoni et al. 2016; Weghorn and Sunyaev 2017; Zhou et al. 2017; Bailey et al. 2018; Martinez-Jimenez et al. 2020).

The new $d_N/d_S - H$ test may be useful to study the selection modes in cancer somatic mutations. Referred to as the $C_N/C_S - H$ test, one difference is that the H-measure can be calculated directly from the cancer somatic mutation data. For instance, the

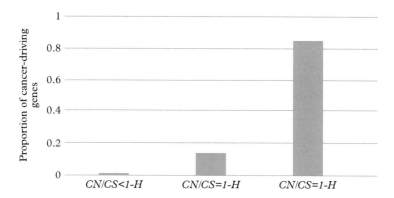

Figure 3.3 *The cancer-version of the d_N/d_S-H, denoted by the C_N/C_S-H test, is applied 294 cancer-driving genes to study selection modes in cancer somatic mutations. A gene belongs to the category of $C_N/C_S = 1$-H is determined by the p-value>0.05 under the null hypothesis; otherwise to the category of $C_N/C_S < 1$-H or $C_N/C_S > 1$-H accordingly.*
Material from: Gu, dN/dS-H, a New Test to Distinguish Different Selection Modes in Protein Evolution and Cancer Evolution, Journal of Molecular Evolution, published 2022, Springer Nature

site-specific number (x) of cancer somatic mutations of a gene can be found in The Cancer Genome Atlas (TCGA) tallied from over ten thousands of patient samples. It is straightforward to calculate the mean and variance of x, and H will be estimated by Eq. (3.13). While an extensive analysis will be published elsewhere, the potential of this approach can be illustrated by an example of 294 cancer-driving genes (Zhao et al. 2021), where the data of cancer somatic mutations were from TCGA. Zhou et al. (2017) showed $C_N/C_S > 1$ in $\sim 70\%$ of those cancer-driving genes, but only a few of them are statistically significant ($p < 0.05$). Impressively, the $C_N/C_S - H$ test revealed that almost all genes (except one) under study have $C_N/C_S > 1 - H$; and $\sim 86\%$ of them are statistically significant ($p < 0.05$) (Fig. 3.3). This preliminary analysis supported the notion that cancer-driving mutations confer a selective advantage on cancer cells (positive selection), with a certain level of purifying selection for the maintenance of protein functionality (Weghorn and Sunyaev 2017; Zhou et al. 2017).

References

Bailey, M.H., Tokheim, C., Porta-Pardo, E., et al. (2018). Comprehensive characterization of cancer driver genes and mutations. *Cell* 173, 371–385

Berglund, A.C., Wallner, B., Elofsson, A., Liberles, D.A. (2005). Tertiary windowing to detect positive diversifying selection. *Journal of Molecular Evolution* 60, 499–504.

Dees, N.D., Zhang, Q., Kandoth, C., et al. (2012). MuSiC: Identifying mutational significance in cancer genomes. *Genome Research* 22, 1589–1598.

Ding, L., Bailey, M.H., Porta-Pardo, E., et al. (2018). Cancer Genome Atlas Research Network. Perspective on Oncogenic Processes at the End of the Beginning of Cancer Genomics. *Cell* 173, 305–320.e10.

Goldman, N., and Yang, Z.H. (1994). Codon-based model of nucleotide substitution for protein-coding dna-sequences. *Molecular Biology and Evolution* 11, 725–736

Gu, X. (2022). d_N/d_S-H, a new test to distinguish different selection modes in protein evolution and cancer evolution. *Journal of Molecular Evolution* 90, 342–351

Gu, X., and Zhang, J. (1997). A simple method for estimating the parameter of substitution rate variation among sites. *Molecular Biology and Evolution* 14, 1106–1113.

Kimura, M. (1983). *The Neutral Theory of Molecular Evolution* (Cambridge University Press, Cambridge, UK; New York).

Kimura, M. (1962). On probability of fixation of mutant genes in a population. *Genetics* 47, 713–719.

Lawrence, M.S., Stojanov, P., Polak, P., et al. (2013). Mutational heterogeneity in cancer and the search for new cancer-associated genes. *Nature* 499, 214–218.

Li, W.H. (1993). Unbiased estimation of the rates of synonymous and nonsynonymous substitution. *Journal of Molecular Evolution* 36, 96–99.

Li, W.H., Wu, C.I., and Luo, C.C. (1985). A new method for estimating synonymous and nonsynonymous rates of nucleotide substitution considering the relative likelihood of nucleotide and codon changes. *Molecular Biology and Evolution* 2, 150–174.

Martinez-Jimenez, F., Muinos, F., Sentis, I., et al. (2020). A compendium of mutational cancer driver genes. *Nature Reviews Cancer* 20, 555–572. https://doi.org/10.1038/s41568-020-0290-x

Mularoni, L., Sabarinathan, R., Deu-Pons, J., et al. (2016). OncodriveFML: a general framework to identify coding and non-coding regions with cancer driver mutations. *Genome Biology* 17:128.

Muse S V, Gaut B S (1994). A likelihood approach for comparing synonymous and nonsynonymous nucleotide substitution rates, with application to the chloroplast genome. *Molecular Biology and Evolution* 11, 715–724

Nei, M., and Gojobori, T. (1986). Simple methods for estimating the numbers of synonymous and nonsynonymous nucleotide substitutions. *Molecular Biology and Evolution* 3, 418–426

Nielsen, R. (2005). Molecular signatures of natural selection. *Annual Review of Genetics* 39, 197–218.

Porta-Pardo, E., and Godzik, A. (2014). e-Driver: a novel method to identify protein regions driving cancer. *Bioinformatics* 30, 3109–3114.

Reimand, J., Wagih, O., and Bader, G. (2013). The mutational landscape of phosphorylation signaling in cancer. *Scientific Reports* 3, 2651.

Stratton, M., Campbell, P., and Futreal, P. (2009). The cancer genome. *Nature* 458, 719–724.

Vogelstein, B., Papadopoulos, N., Velculescu, V.E., et al. (2013). Cancer genome landscapes. *Science* 339, 1546–1558.

Weghorn, D., and Sunyaev, S. (2017). Bayesian inference of negative and positive selection in human cancers. *Nature Genetics* 49:1785–1788.

Yang, Z. (1997). PAML: a program package for phylogenetic analysis by maximum likelihood. *Computer Applications in the Biosciences* 13, 555–556.

Zhao W, Yang J, Wu J, et al. (2021). CanDriS: Posterior profiling of cancer-driving sites based on two-component evolutionary model. *Briefings in Bioinformatics* 22: bbab131.

Zhou, Z., Zou, Y., Liu, G., et al. (2017). Mutation-profile-based methods for understanding selection forces in cancer somatic mutations: A comparative analysis. *Oncotarget* 8, 58835–58846.

4

Functional Divergence after Gene Duplication
DIVERGE Analysis

Many organisms have undergone genome-wide or local chromosome duplication events during their evolution (Ohno 1970; Holland e al. 1994; Wolfe and Shields 1997; Gu and Nei 1999; Dermitzakis and Clark 2001; Gu et al. 2002b). As a result, many genes are represented as several paralogs in the genome with related but distinct functions (gene families). Since gene duplication is thought to have provided the raw materials for functional innovations, it is desirable to identify amino acid sites that are responsible for functional divergence from the sequence analysis of a gene family. Several computational methods have been proposed (e.g., Casari et al. 1995; Lichtarge et al. 1996; Livingstone and Barton 1996; Landgraf et al. 1999). As most amino acid substitutions are not related to the functional divergence but only represent neutral evolution, it becomes crucial how to statistically distinguish between these two possibilities. To this end, Gu (1999, 2001a, 2001b, 2006)has developed a series of statistical models, based on the principle that functional divergence between duplicate genes is highly correlated with the change of evolutionary rate after the gene duplication. In this chapter, we discuss these statistical and computational methods. Case-studies for applying these methods in biological problems will be presented in the next chapter.

4.1 Modeling functional divergence

Consider a multiple sequence alignment (MSA) of a gene family with two duplicate gene clusters 1 and 2, respectively (Fig. 4.1). Although various terminologies were used in previous literature, the evolutionary patterns of amino acid sites of a gene family can be concisely classified into the following types:

Statistical Analysis of Molecular and Genomic Evolution. Xun Gu, Oxford University Press. © Xun Gu (2024).
DOI: 10.1093/oso/9780198816515.003.0004

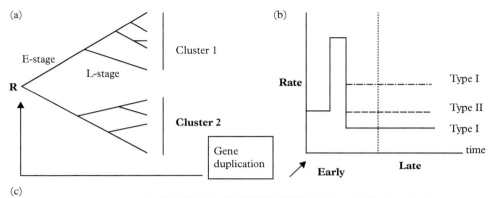

	Sequence	Type 0	Type 1	Type 2	Type-U
Gene 1	1	CR	WQLV	RV	KTLI
	2	CR	WQIV	RV	RVLL
	3	CR	WQVG	RV	KIIV
	4	CR	WQVG	RV	NVLL
	5	CR	WQAT	RV	DMLL
	6	CR	WQAT	RV	IKIL
	7	CR	WQVI	RV	EKLI
	8	CR	WQIT	RV	DLVL
Gene 2	9	CR	LTFD	DR	LKLM
	10	CR	ITFD	DR	QLVV
	11	CR	ITFD	ER	RLVV
	12	CR	YSFD	DK	LHVV
	13	CR	LEFD	DR	KMAL
	14	CR	LEFE	DR	KLLI
	15	CR	LEFD	DR	KLLL
	16	CR	VGFD	DK	ELII
	17	CR	VTFD	DR	RLII

Figure 4.1 *(A) Two gene clusters after gene duplication. E and L are early and late stages of gene clusters 1 and 2, respectively. (B) Type I and type II functional divergences after gene duplication. In the early stage, the evolutionary rate (say, in cluster 1) may increase for functional divergence-related change, but in the late stage it may be higher (or lower) than its original rate, resulting in shifted functional constraints between clusters 1 and 2, or type-I functional divergence. If the rate in the late stage is back to the same as the original one, no shifted functional constraints between clusters 1 and 2 can be observed, or type-II functional divergence. (C) A hypothetical multiple-alignment to show universally conserved sites (type 0), type I, and type II amino acid patterns, U-type sites (unclassified).*

Reproduced with permission from X Gu, Statistical methods for testing functional divergence after gene duplication., Molecular Biology and Evolution, Volume 16, Issue 12, 1 December 1999, Pages 1664–1674, https://doi.org/10.1093/oxfordjournals.molbev.a026080

(*i*) Type 0: it represents amino acid patterns that are universally conserved through the whole gene family, implying that these sites are important for the common function shared by duplicate genes.

(*ii*) Type I: it represents amino acid patterns that are very conserved in duplicate gene cluster 1 but highly variable in cluster 2, or *vice versa*, implying that these sites may have experienced shifted functional constraints.

(*iii*) Type II: it represents amino acid patterns that are highly conserved in both duplicate gene clusters but their biochemical properties are distinct, e.g., charge positive versus negative, implying that these sites may have been responsible for functional specifications.

(*iv*) Type U: it represents amino acid patterns that cannot be unambiguously classified into any three above types, such as amino acid sites that are highly variable in both clusters.

Based on the argument that functional divergence between duplicates tended to occur in the early stage after gene duplication, we further formulate two basic mechanisms as follows.

(*i*) *Type-I functional divergence*: this postulates that functional divergence between duplicate genes results in shifted functional constraints (i.e., different evolutionary rates) at some amino acid sites, or the type-I pattern. If a site is related to type-I functional divergence, known as an F_1-site, the evolutionary rate at this site is assumed to be independent between duplicate clusters. In other words, functional constraint of an F_1-site in one duplicate gene contains no information to predict the evolutionary conservation at the other one.

(*ii*) *Type-II functional divergence*: this postulates that functional divergence between duplicate genes at some sites results in a type-II pattern. If a site is related to type-II functional divergence, known as an F_2-site, we assume that the evolutionary rate at this site may have been decoupled between the early and late stages after gene duplication. Though radical amino acid changes occur frequently in type-II functional divergence, both duplicate genes remain virtually the same, with a high level of sequence conservation.

In the following we discuss several statistical methods that have been well developed for detecting functional divergence between duplicate genes. In particular, we address two important issues: Whether the functional divergence between duplicate genes is statistically significant? And if it is the case, how can we predict amino acid sites that are largely responsible for these functional divergences?

4.2   Statistical methods for type I functional divergence

4.2.1   The two-state (F_0/F_1) model

Suppose that a gene family contains two duplicate gene clusters (Fig. 4.1). Orthologous genes within each cluster are functionally equivalent, suggesting that the evolutionary rate of a site remains virtually constant though it may vary among different sites. As a molecular clock is not assumed, lineage-specific factors such as generation-time effects will not affect our analysis. Without loss of generality, the evolutionary rates in gene cluster 1 and gene cluster 2 are simply denoted by λ_1 and λ_2, respectively.

Suppose we already know exactly which sites are related to type-I functional divergence so that all sites can be classified into either of two states:

(*i*) F_0 (functional divergence-unrelated): The evolutionary rate (λ) of an F_0-site is the same between duplicate gene clusters, indicating no shift in functional constraint, that is, $\lambda_1 = \lambda_2 = \lambda$.

(*ii*) F_1 (type-I functional divergence-related): The evolutionary rate of an F_1-site has no correlation between duplicate gene clusters, as such a site has experienced the shifted functional constraint. Consequently, λ_1 and λ_2 are statistically independent.

In practice, however, we do not know to which state each site belongs. This problem can be solved by implementing a two-state mixture model: an amino acid site is in the state of F_1 with a probability of $P(F_1)$, or the state of F_0 with a probability of $P(F_0)$; obviously $P(F_1) + P(F_0) = 1$. Denote $\theta_I = P(F_1)$, called *the coefficient of type-I functional divergence*. As θ_I increases from 0 to 1, the functional divergence between two clusters increases from very weak to extremely strong. Next we formulate a statistical test to evaluate the significance of type-I functional divergence after gene duplication, e.g., the null hypothesis $\theta_I = 0$ versus the alternative $\theta_I > 0$. To this end, a statistical model for the gene family evolution should be developed at first.

4.2.2   The Poisson-gamma model of protein sequence evolution

Amino acid substitutions

A simple model for protein sequence evolution is the Poisson process. At a given site, the number of amino acid changes in each cluster, denoted by X_1 and X_2 for gene clusters 1 and 2, respectively, follows a Poisson distribution. That is, the probability of X_i changes is given by

$$p(X_i|\lambda_i) = \frac{(\lambda_i T_i)^{X_i}}{X_i!} e^{-\lambda_i T_i}, \quad i = 1, 2 \tag{4.1}$$

where T_1 and T_2 are the total evolutionary times of clusters 1 and 2, respectively.

Corrected number of minimum-required changes

To apply Eq. (4.1), one needs to know the number of changes at each site in each gene cluster. Since these numbers (X_1 and X_2) cannot be observed directly from the multiple alignments, a conventional solution is to use the number of minimum-required changes (m) as an approximation, which can be inferred by the parsimony under a known phylogenetic tree (Fitch 1971). However, m is a biased "estimate" for the true number of changes because it does not consider the possibility of multiple-hits (Wakeley 1993). This parsimonious bias can be corrected by the method developed by Gu and Zhang (1997). Under a known phylogeny, they showed that the expected number of changes ($X = X_1$ or X_2) at a given site is the non-negative solution of the likelihood equation

$$\sum_{i=1}^{M} \frac{\delta_i b_i}{1 - e^{-\hat{X}b_i/B}} = 1 \tag{4.2}$$

where B is the total branch length of the gene cluster, b_i is the i-th branch length, $i = 1, \ldots, M$ (M is the total number of branches); $\delta_i = 1$ if there is an amino acid change in the i-th branch, otherwise $\delta_i = 0$. Computer simulations showed that the estimate of the corrected number of changes was asymptotically unbiased. Here we mention two special cases that may be useful in practice: (1) $\hat{X} \approx m$ for short branch lengths, and (2) $\hat{X} = -M\ln(1 - m/M)$ for equal branch lengths.

Variation of evolutionary rate over sites

Because amino acid sites differ in functional constraints, we expect that the evolutionary rate (λ) varies among amino acid sites (Kimura 1983). Since the rate λ at each site is usually unknown, a common practice is to assume that the rate variation among sites follows to a specified distribution (Yang 1993; Gu et al. 1995), e.g., the widely-used gamma distribution

$$\phi(\lambda) = \frac{\beta^{\alpha}}{\Gamma(\alpha)} \lambda^{\alpha-1} e^{-\beta\lambda} \tag{4.3}$$

where $\lambda = \lambda_1$ or λ_2, respectively. The shape parameter α describes the degree of rate variation among sites, whereas β is a scalar. Since $1/\sqrt{\alpha}$ is the coefficient of variation of λ, the larger the α value is, the weaker the rate variation is, and $\alpha = \infty$ means a uniform rate among sites.

4.2.3   Estimation of type-I functional divergence

Building the likelihood function

Gu (1999) formulated $P(X_1, X_2)$, the joint distribution of the numbers of amino acid changes in clusters 1 and 2, as follows. Since the two gene clusters are monophyletic (Fig. 4.1), it is reasonable to assume that two Poisson processes of amino acid substitutions at this site, $p(X_1|\lambda_1)$ and $p(X_2|\lambda_2)$, are independent, and so the joint distribution of X_1 and X_2 is given by $P(X_1, X_2|\lambda_1, \lambda_2) = p(X_1|\lambda_1)p(X_2|\lambda_2)$. The next step is to integrate the evolutionary rates λ_1 and λ_2 out, both of which vary among sites according to the gamma distribution $\phi(\lambda)$ in Eq. (4.3).

Likelihood function under F_0

Consider the state of F_0, under which the site is functional divergence-unrelated, implying virtually the same evolutionary rate $\lambda_1 = \lambda_2 = \lambda$. Therefore, the joint distribution of X_1 and X_2 under F_0 is given by

$$P(X_1, X_2|F_0) = \int_0^\infty p(X_1|\lambda)p(X_2|\lambda)\phi(\lambda)d\lambda \tag{4.4}$$

Let D_1 and D_2 be the expected number of amino acid changes per site in duplicate clusters 1 and 2, respectively. From Eq. (4.1) to Eq. (4.3), Gu (1999) derived $P(X_1, X_2|F_0) = K_{12}(i,j)$ for $X_1 = i$ and $X_2 = j$, where

$$K_{12}(i,j) = \frac{\Gamma(i+j+\alpha)}{i!j!\Gamma(\alpha)} \left(\frac{D_1}{D_1 + D_2 + \alpha}\right)^i \left(\frac{D_2}{D_1 + D_2 + \alpha}\right)^j \left(\frac{\alpha}{D_1 + D_2 + \alpha}\right)^\alpha \tag{4.5}$$

Likelihood function under F_1

Consider the state of F_1, under which the amino acid site is functional divergence-related, implying λ_1 and λ_2 are independent. This immediately results in the joint distribution of X_1 and X_2 under F_1 as $P(X_1, X_2|F_1) = P(X_1|F_1) \times P(X_2|F_1)$. For each cluster, the distribution of the number of changes $(X = X_1$ or $X_2)$ under F_1 is given by

$$P(X|F_1) = \int_0^\infty p(X|\lambda)\phi(\lambda)d\lambda \tag{4.6}$$

For $X_1 = i$ one can show $P(X_1|F_1) = Q_1(i)$ where

$$Q_1(i) = \frac{\Gamma(i+\alpha)}{i!\Gamma(\alpha)} \left(\frac{D_1}{D_1 + \alpha}\right)^i \left(\frac{\alpha}{D_1 + \alpha}\right)^\alpha \tag{4.7}$$

For $X_2 = j$ we have $P(X_2|F_1) = Q_2(j)$, where $Q_2(j)$ is the same form of $Q_1(i)$ except that D_1 replaced by D_2.

Likelihood function and estimation

Putting all the above together, under the two-state model of functional divergence, one can write the joint distribution of X_1 and X_2 as $P(X_1, X_2) = P(F_0)P(X_1, X_2|F_0) + P(F_1)P(X_1, X_2|F_1)$. Since $P(F_1) = \theta_I$ and $P(F_0) = 1 - \theta_I$, we obtain the analytical result

$$P(X_1, X_2) = (1 - \theta_I)K_{12} + \theta_I Q_1 Q_2 \qquad (4.8)$$

which provides the statistical foundation for estimating type-I functional divergence.

Let $P(X_1 = i_k, X_2 = j_k)$ be the probability of $X_1 = i_k$ and $X_2 = j_k$ at site k. Then, given the observed number of changes in two clusters along the sites, the likelihood function can be written as

$$L = \prod_k P(X_1 = i_k, X_2 = j_k) \qquad (4.9)$$

There are four unknown parameters, D_1, D_2, α, and θ_I, which can be numerically estimated by a standard maximum likelihood approach. Using appropriate initial values, the maximum likelihood (ML) estimates of D_1, D_2, α, and θ_I, as well as their approximate sampling variances, can be obtained numerically.

Apparently, we are mostly interested whether θ_I, the coefficient of type-I functional divergence, is significantly greater than 0, which can be asserted by the likelihood ratio test (LRT): The null hypothesis is $H_0 : \theta_I = 0$ versus the alternative $H_A : \theta_I > 0$. Based on the likelihood ratio LR, it is known that the statistic $\delta = -2\ln(LR)$ asymptotically follows a $\chi^2_{[1]}$ distribution with one degree of freedom.

4.2.4 Predicting critical amino acid residues

If the likelihood ratio test provides strong statistical evidence for the type-I functional divergence after gene duplication (i.e., $\theta_I > 0$), it is of great interest to predict which sites are likely to be responsible for these functional differences. Gu (1999) has addressed this issue, developing a site-specific profile to calculate the posterior probability that an amino acid site is functional divergence-related, given the observed numbers of amino acid changes in two duplicate clusters.

Under the two-state model of functional divergence, each site has two possible states, F_0 and F_1. From the Bayesian view, $P(F_1)$ and $P(F_0)$ are treated as priors. In other words, the coefficient of type-I functional divergence $\theta_I = P(F_1)$ is interpreted as the prior probability of the state of being functional divergence-related. Therefore, to provide a statistical basis for predicting which state (F_0 or F_1) is more likely at a given site, we need to compute the posterior probability of state F_1 at this site with X_1 and X_2 changes in clusters 1 (and 2), respectively, denoted by $P(F_1|X_1, X_2)$. According to the Bayesian law, we have

$$P(F_1|X_1, X_2) = \frac{P(F_1)P(X_1, X_2|F_1)}{P(X_1, X_2)} = \frac{\theta_I Q_1 Q_2}{(1 - \theta_I)K_{12} + \theta_I Q_1 Q_2} \tag{4.10}$$

and $P(F_0|X_1, X_2) = 1 - P(F_1|X_1, X_2)$. To be concise, let q_k be the posterior probability $P(F_1|X_1, X_2)$ at site k, $k = 1, \ldots, L$, where L is the number of sites. From Eq. (4.10), one can further show the following relationship

$$\theta_I = \sum_{X_1, X_2} P(X_1, X_2) \times P(F_1|X_1, X_2) \approx \sum_{k=1}^{L} q_k/L \tag{4.11}$$

Hence, those probabilities for single sites being functional divergence-related can be viewed as a (weighted) spectrum of the coefficient of type-I functional divergence. Approximately, θ_I is the average of posterior probabilities over all sites.

4.2.5   Reduced rate correlation between duplicate genes: an alternative view of θ_I

Consider the case of two duplicate gene clusters (Fig. 4.1). If all amino acid sites have experienced no functional divergence after gene duplication, the two duplicate genes have no shifted functional constraints so that the evolutionary rates (λ_1 and λ_2) are virtually equal over all sites. In other words, the coefficient of correlation between λ_1 and λ_2, denoted by r_λ, could be very close to 1. Obviously, shifted functional constraints at some sites can result in different evolutionary rates, reducing the rate correlation between them, i.e., $r_\lambda < 1$. Hence, the shifted functional constraints between two gene clusters, or type-I functional divergence, can be measured by r_λ, the coefficient of rate correlation between λ_1 and λ_2. Let D_1 and V_1 (or D_2 and V_2) be the mean and variance of the number (X_1 or X_2) of changes (over sites) in cluster 1 (or cluster 2), respectively, and σ_{12} be the covariance (over sites) between them. Under the Poisson-gamma model of amino acid substitutions, the variances of λ_1 and λ_2 are given by $Var(\lambda_1) = V_1 - D_1$ and $Var(\lambda_2) = V_2 - D_2$, respectively, and the covariance $Cov(\lambda_1, \lambda_2) = \sigma_{12}$. Therefore, r_λ can be calculated by the following formula

$$r_\lambda = \frac{\sigma_{12}}{\sqrt{(V_1 - D_1)(V_2 - D_2)}} \tag{4.12}$$

Under the two-state model of type-I functional divergence, it is easy to show that the variances of λ_1 and λ_2 are the same, i.e., $Var(\lambda_1) = Var(\lambda_2) = Var(\lambda)$, and the covariance between λ_1 and λ_2 is given by $Cov(\lambda_1, \lambda_2) = (1 - \theta_I)Var(\lambda)$. This argument directly leads to the following simple relationship between r_λ and θ_I, that is,

$$r_\lambda = 1 - \theta_I \qquad (4.13)$$

Therefore, the coefficient of type-I functional divergence can also interpreted as the reduced coefficient of rate correlation between duplicate clusters. Moreover, it provides a "model-free" approach to estimate θ_I by $1 - r_\lambda$, because it does not rely on a specified model for the functional divergence. Gu (1999) derived an approximate formula for the sampling variance

$$Var(\hat{\theta}_\lambda) \approx \frac{1}{L-3} \left(\frac{1 - r_X^2}{r_M} \right)^2 \qquad (4.14)$$

where L is number of sites, $r_X = \sigma_{12}/\sqrt{V_1 V_2}$ and $r_M = \sqrt{(1 - D_1/V_1)(1 - D_2/V_2)}$.

4.3 Statistical methods for type-II functional divergence

4.3.1 Modeling type-II functional divergence

Early and late stages after gene duplication

In principle, the evolution of protein sequences of duplicate genes can be divided into two stages, the early (E) stage after gene duplication, and the late (L) stage (Fig. 4.1). It has been assumed that type-II functional divergence between duplicate genes has occurred in the E-stage, while in the L-stage, the purifying selection plays a major role to maintain related but distinct functions of two duplicate genes. Accordingly, Gu (2006) specified a two-state model for type-II functional divergence:

(*i*) In the E-stage, an amino acid site can be in either of two states: F_0 (functional divergence-unrelated) and F_2 (type-II functional divergence-related). The probability of a site being under F_2 is $P(F_2) = \theta_{II}$, called *the coefficient of type-II functional divergence*.

(*ii*) In the L-stage, an amino acid site is always under the state of F_0, indicating no further functional divergence. Amino acid substitutions in this stage are mainly under the purifying selection.

Substitution models under F_0 and F_2

During the evolution the pattern of amino acid substitutions, or the substitution model, relies on the states of functional divergence (F_0/F_2). The F_0-substitution model largely reflects the conserved evolution of protein sequences, which can be empirically determined by the Dayhoff model (Dayhoff 1978), or the JTT model (Jones et al. 1992). By contrast, under F_2, radical amino acid substitutions may occur more frequently, apparently due to the functional divergence between duplicate genes (Lichtarge et al. 1996). To avoid over-parameterization, Gu (2006) proposed a simple substitution model that can distinguish between the *radical* and *conserved* amino acid substitutions.

(*i*) Classify 20 amino acids into 4 groups: charge positive (K, R, H), charge negative (D, E), hydrophilic (S, T, N, Q, C, G, P), and hydrophobic (A, I, L, M, F, W, V, Y). An amino acid substitution is called radical (denoted by **R**) if it changes from one group to another; otherwise it is called conserved, i.e., within the group, denoted by **C**. The status of no substitution is denoted by **N**.

(*ii*) Under the state of F_0, the transition probability for a radical, conserved, or no substitution, follows an extended Poisson model, that is,

$$P(\mathbf{R}|F_0) = \pi_R(1 - e^{-\lambda t})$$
$$P(\mathbf{C}|F_0) = \pi_C(1 - e^{-\lambda t})$$
$$P(\mathbf{N}|F_0) = e^{-\lambda t} \tag{4.15}$$

respectively, where t is the evolutionary time, λ is the substitution rate, and π_R (π_C) is the proportion of radical (or conserved) substitutions in the total substitutions; $\pi_R + \pi_C = 1$. Based on the Dayhoff PAM matrix, one can empirically determine $\pi_R = 0.312$ and $\pi_C = 0.688$. Indeed, without any functional divergence, conserved amino acid substitutions are more likely to occur, as expected by the theory of neutral evolution (Kimura 1983).

(*iii*) Consider the transition probabilities under F_2 in the early stage, denoted by $P(Y|F_2)$ for $Y = \mathbf{N}, \mathbf{R}, \mathbf{C}$. By definition, an amino acid site that has no change in the early stage is essentially unrelated to the type-II functional divergence, implying that $P(\mathbf{N}|F_0) = 0$. Further, after assuming that the process of functional divergence is a fast process, we obtain

$$P(\mathbf{R}|F_2) = a_R$$
$$P(\mathbf{C}|F_2) = a_C$$
$$P(\mathbf{N}|F_2) = 0 \tag{4.16}$$

That is, a_R (or a_C) is the proportion of F_2-radical (or conserved) substitutions. Note that the F_2-radical amino acid substitution (a_R) can be much higher than that under F_0 (π_R).

Evolutionary link between early and late stages

The evolutionary link between early and late stages depends on the status of type-II functional divergence. Let λ_E and λ_L be the evolutionary rates in the E and L stages, respectively. The statistical framework is under the following assumptions:

(*i*) A random variable u, called the rate component, varies among sites according to a standard gamma distribution, where the shape parameter α describes the strength of rate variation among sites (Gu et al. 1995).

(*ii*) Under F_0, the evolutionary rates in the early (λ_E) and late (λ_L) stages share the same rate component u. That is, $\lambda_E = u$ and $\lambda_L = u$.

(*iii*) F_2-amino acid substitutions in the early stage is independent of the rate component u. In other words, F_2-amino acid substitutions have escaped from the ancestral functional constraint on the protein sequence.

4.3.2 Two duplicate gene clusters

Consider two phylogenetic clusters generated by gene duplication, each of which consists of several orthologous genes (Fig. 4.1). Let X be the amino acid pattern of the L-stage, a column of amino acid sites in the multiple alignment of the sequences. Let $Y = (a, b)$ be the amino acid pattern of the E-stage, the ancestral sequences of two internal nodes a and b. From the above assumption (*ii*), the joint probability of X and Y under F_0 is given by

$$P(X, Y|F_0) = \int_0^\infty P(X|Y)P(Y|F_0)\phi(u)\,du \qquad (4.17)$$

where $P(Y|F_0)$ is determined by Eq. (4.15) for $Y = \mathbf{N}, \mathbf{C}$, or $\mathbf{R}$, respectively, and $P(X|Y)$ is the likelihood of the subtrees of two clusters A and B, conditional on the ancestral states a and b, (Gu 2001b). Similarly, from the above assumption (*iii*), under F_2 we have

$$P(X, Y|F_2) = P(Y|F_2) \times \int_0^\infty P(X|Y)\phi(u)\,du \qquad (4.18)$$

where $P(Y|F_2)$ is given by Eq. (4.16). Recalling that the probability of a site being under F_2 is given by $P(F_2) = \theta_{II}$, the coefficient of type-II functional divergence, we have the joint probability for X and Y as follows

$$P(X, Y) = (1 - \theta_{II})P(X, Y|F_0) + \theta_{II}P(X, Y|F_2) \qquad (4.19)$$

4.3.3 Building the likelihood function for estimating θ_{II}

Direct application of Eq. (4.19) for estimating θ_{II} is difficult because the amino acid pattern of early-stage (Y) is unobservable. One may invoke the ancestral sequence inference (Yang et al. 1995) to solve this problem, treating the inferred ancestral sequences as observations. However, this approach requires extensive computations and is sensitive to the statistical uncertainty in ancestral sequence inference. Nevertheless, Gu (2006) proposed a simple, robust method that is computationally efficient.

The Poisson-based model of $P(X|Y)$

Testing type-II functional divergence in the early stage between two gene clusters utilizes the within-cluster amino acid patterns to evaluate the degree of conservation in the late-stage. Gu (2006) implemented a Poisson-based model to derive $P(X|Y)$, the probability of the late-stage (X), conditional of the early-stage (Y), which counts the number (k) of substitutions. Indeed, smaller values of k of substitutions in a gene cluster indicate high conservation, and *vice versa*. Formally, at a given amino acid site, the number of substitutions in each cluster $(A$ or $B)$ follows a Poisson process, e.g., for cluster A, we have

$$p_A(k) = \frac{(\lambda_A T_A)^k}{k!} e^{-\lambda_A T_A} \tag{4.20}$$

with the same applying to $p_B(k)$, where T_A (or T_B) is the total evolutionary time of cluster A (or B), and λ_A (or λ_B) is the evolutionary rate of cluster A (or B), respectively. Thus, $P(X|Y)$ is specifically written as $P(X = (i,j)|Y)$ under the Poisson model, where i or j is the number of substitutions in cluster A or B. Note that the Poisson property implies that the process of late-stage (X) is independent of the early stage Y, resulting in

$$P[X = (i,j)|Y] = p_A(i)p_B(j) \tag{4.21}$$

Early-late joint distribution conditional of F_0

Rewrite Eq. (4.17) under the Poisson-based model as follows

$$f_{ij,Y|F_0} = P[X = (i,j), Y|F_0] = \int_0^\infty p_A(i)p_B(j)P(Y|F_0)\phi(u)du \tag{4.22}$$

where $\phi(u)$ is a gamma distribution. Under F_0, the evolutionary rates in the early and the late stages share the same rate component u, that is, $\lambda_A = \lambda_B = u$ in Eq. (4.20) and $\lambda = u$ in Eq. (4.15); here λ refers to the rate in the early stage. After some algebra, the analytical form of $f_{ij,Y|F_0}$ can be derived. Define two sets of new variables. The first set is

$$Z = \alpha/(D_A + D_B + \alpha)$$

$$Z_A = D_A/(D_A + D_B + \alpha)$$

$$Z_B = D_B/(D_A + D_B + \alpha) \tag{4.23}$$

where $D_A = uT_A$ and $D_B = uT_B$ are the total branch lengths of clusters A and B, respectively, and α is the shape parameter. The second set of new variables is similarly defined by

$$W = \alpha/(D_A + D_B + d + \alpha)$$

$$W_A = D_A/(D_A + D_B + d + \alpha)$$

$$W_B = D_B/(D_A + D_B + d + \alpha) \tag{4.24}$$

where $d = ut$ is the branch length of the early stage. We then calculate

$$Q_{ij} = \int_0^\infty p_A(i)p_B(j)\phi(u)\,du$$

$$= \frac{\Gamma(i+j+\alpha)}{i!\,j!\,\Gamma(\alpha)}D^\alpha D_A^i D_B^j \tag{4.25}$$

and

$$M_{ij} = \int_0^\infty e^{-ut}p_A(i)p_B(j)\phi(u)\,du$$

$$= \frac{\Gamma(i+j+\alpha)}{i!\,j!\,\Gamma(\alpha)}W^\alpha W_A^i W_B^j \tag{4.26}$$

respectively. Putting these together, one can show

$$f_{ij,N|F_0} = M_{ij}$$

$$f_{ij,R|F_0} = (Q_{ij} - M_{ij})\,\pi_R$$

$$f_{ij,C|F_0} = (Q_{ij} - M_{ij})\,\pi_C \tag{4.27}$$

Early-late joint distribution conditional of F_2

One can derive the early-late conditional of F_2 in the same vein. Rewrite Eq. (4.18) under the Poisson-based model as follows

$$f_{ij,Y|F_2} = P[X = (i,j),\,Y|F_0] = P(Y|F_2)\int_0^\infty p_A(i)p_B(j)\phi(u)\,du \tag{4.28}$$

Together with Eq. (4.16), we have

$$f_{ij,N|F_2} = 0$$
$$f_{ij,R|F_2} = Q_{ij}a_R$$
$$f_{ij,C|F_2} = Q_{ij}a_C \tag{4.29}$$

respectively.

Joint early-late joint distribution and ML estimation

It follows that the analytical form of early-late joint distribution under the Poisson model, $f_{ij,Y} = P[X = (i,j), Y]$ for $Y = \mathbf{N}, \mathbf{R}, \mathbf{C}$, can be written as follows

$$f_{ij,N} = (1 - \theta_{II})M_{ij}$$
$$f_{ij,R} = (1 - \theta_{II})(Q_{ij} - M_{ij})\pi_R + \theta_{II}\,a_R Q_{ij}$$
$$f_{ij,C} = (1 - \theta_{II})(Q_{ij} - M_{ij})\pi_C + \theta_{II}\,a_C Q_{ij} \tag{4.30}$$

Based on the model of $f_{ij,Y}$, a ML can be formulated to estimate θ_{II}. Let $n_{ij,Y}$ be the number of sites with the late-stage pattern $X = (i,j)$ and the early stage pattern $Y = \mathbf{N}, \mathbf{Y}, \mathbf{C}$. Thus, the likelihood function can be written as

$$L = \prod_{i,j,Y} f_{ij,Y}^{n_{ij,Y}} \tag{4.31}$$

   Gu (2006) implemented a practically feasible algorithm to estimate unknown parameters when the phylogenetic tree of the gene family is known or can be reliably inferred. Note that the distribution of late stage, i.e., the probability of a site being i and j substitutions in the two clusters, is given by

$$P[X = (i,j)] = f_{ij,N} + f_{ij,C} + f_{ij,R} = Q_{ij} \tag{4.32}$$

which depends on three (late-stage) parameters D_A, D_B, and α. Hence, the method of Gu and Zhang (1997) can be used to obtain the ML estimates of these parameters. After replacing three unknown late-stage parameters by their ML estimates, a likelihood approach was developed to estimate the early-stage parameters θ_{II}, a_R/a_C, and d, respectively, based on the inferred ancestral sequences of the early stage.

U-likelihood estimation of θ_{II}

A simplified method called the U-likelihood has been shown useful, which utilizes amino acid sites that are universally conserved in both clusters, i.e., $i = j = 0$. Let n_{00Y} be the number of sites with $Y = \mathbf{N}, \mathbf{R}, \mathbf{C}$, respectively. Let $n_{00} = n_{00N} + n_{00R} + n_{00C}$, and $f_{00} = f_{00N} + f_{00R} + f_{00C}$. Then, the U-likelihood can be built as follows

$$L = (1 - f_{00})^{N - n_{00}} \times \prod_{Y=N,R,C} f_{00,Y}^{n_{00,Y}} \tag{4.33}$$

Let $\hat{f}_{00N} = n_{00N}/N$. It has been shown that the U-ML estimates of θ_{II} and d are given by

$$\theta_{II} = 1 - \hat{f}_{00,N} \left[1 + \frac{\hat{D}_A + \hat{D}_B + d}{\hat{\alpha}} \right]^{\hat{\alpha}}$$

$$d = -\ln(1 - p) + \ln(1 - \theta_{II}) \tag{4.34}$$

where p is the proportion of different amino acid sites between the ancestral nodes of two clusters. Since the U-method largely relies on the universally conserved sites, it seems robust against the inaccuracy of ancestral sequence inference and sequence alignment.

4.3.4 Predicting critical amino acid residues: empirical Bayesian approach

The identification of which sites are responsible for these type II functional differences is of great interest, if the coefficient of functional divergence (θ_{II}) between early and late stages is significantly larger than 0. To this end, we wish to know the posterior probability of state F_2 in the early stage at a site, i.e., $P(F_2|X, Y)$. According to the Bayesian law, we have

$$P(F_2|X, Y) = \frac{P(F_2)P(X, Y|F_2)}{P(X, Y)} \tag{4.35}$$

where the prior probability of F_2 in the early stage is given by $P(F_2) = \theta_{II}$. Under the Poisson-based model, we have $P(X, Y|F_2) = f_{ij,Y|F_2}$ by Eq. (4.29), $P(X, Y|F_0) = f_{ij,Y|F_0}$ by Eq. (4.27), and $P(X, Y) = f_{ij,Y}$ by Eq. (4.30), respectively. Together, one can show

$$P(F_2|X, Y) = 0 \qquad\qquad \text{if} \quad Y = N$$

$$P(F_2|X, Y) = \frac{a_C \theta_{II} Q_{ij}}{(1 - \theta_{II})(Q_{ij} - M_{ij})\pi_C + \theta_{II} a_C Q_{ij}} \qquad \text{if} \quad Y = C$$

$$P(F_2|X, Y) = \frac{a_R \theta_{II} Q_{ij}}{(1 - \theta_{II})(Q_{ij} - M_{ij})\pi_R + \theta_{II} a_R Q_{ij}} \qquad \text{if} \quad Y = R$$

One may find it is convenient to use the posterior probability ratio of F_2 to F_0, i.e., $R(F_2|F_0) = P(F_2|X, Y)/P(F_0|X, Y)$. After some algebra, we obtain

$$R(F_2|F_0) = 0 \qquad\qquad\qquad\qquad\qquad \text{if} \qquad Y = N$$

$$R(F_2|F_0) = \frac{\theta_{II}}{1 - \theta_{II}} \frac{a_C}{\pi_C} \frac{1}{1 - (1 - h)^{i+j+\alpha}} \qquad \text{if} \qquad Y = C$$

$$R(F_2|F_0) = \frac{\theta_{II}}{1 - \theta_{II}} \frac{a_R}{\pi_R} \frac{1}{1 - (1 - h)^{i+j+\alpha}} \qquad \text{if} \qquad Y = R \qquad (4.37)$$

where $h = d/(D_A + D_B + d + \alpha)$.

An important observation is that the posterior ratio $R(F_2|F_0)$ reaches its maximum if there is no amino acid substitution in each gene cluster but the amino acid is different between them, i.e., $i = j = 0$ and $Y \neq N$. Assuming that the proportion of radical changes under F_2 is higher than that under F_0 such that $a_R/a_C > \pi_R/\pi_C$ holds, we have

$$R(F_2|F_0)_{max} = \frac{\theta_{II}}{1 - \theta_{II}} \frac{a_R}{\pi_R} \frac{1}{1 - (1 - h)^{\alpha}} \qquad\qquad (4.38)$$

Hence, a typical cluster-specific site will receive a highest score for the type-II functional divergence, consistent with biological intuition. It should be noticed that a high score could be statistically meaningless if θ_{II} is not significantly larger than 0. Finally, the fact of $R(F_2|F_0)_{max} \to \infty$ when $h \to 0$ indicates that greater accuracy can be achieved as more sequences are analyzed (i.e., increasing D_A or D_B). In practice, one may use this property to determine how many sequences are sufficient to achieve the statistical resolution of F_2-site prediction.

4.4 Functional-structural basis of type-I functional divergence between caspases

A cascade of cysteine aspartyl proteases (caspases) is the key component in the apoptotic machinery (or programmed cell death). There are 14 members of the caspase gene family in mammals, which can be further classified into two major subfamilies, CED-3 (including caspase-2, -3, -6, -7, -8, -9, -10, and -14) and ICE (including caspase-1, -4, -5, -11, -12, and -13). CED-3-type caspases are essential for most apoptotic pathways, while the major function of the ICE-type caspases is to mediate immune response. Phylogenetic analysis showed that these major caspase subfamilies are clustered separately (Fig. 4.2).

Using DIVERGE, Wang and Gu (2001) analyzed the caspase gene family to explore the structural-functional basis for the type-I functional divergence of protein sequences

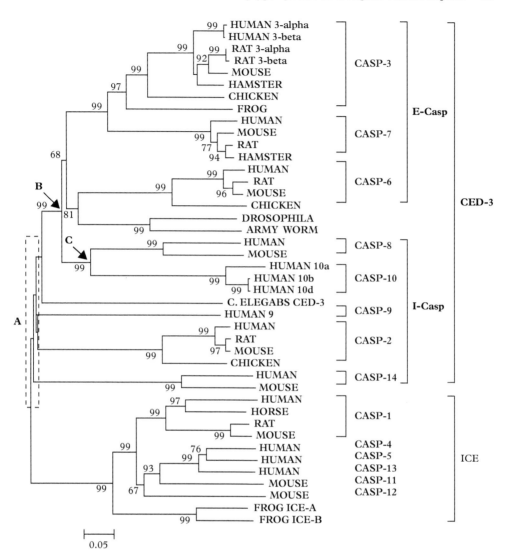

Figure 4.2 *Phylogenetic tree of the caspase gene family, inferred by the neighbor-joining method on the basis of amino acid sequences with the Poisson correction. Bootstrap values >50% are presented. Initiator caspases (I-casps) are involved in the upstream regulatory events, and effector caspases (E-casps) directly lead to cell disassembly.*

between CED-3 and ICE caspase subfamilies. Based on the inferred tree of caspases (Fig. 4.2), we found that type I functional divergence is statistically significant between two major subfamilies, CED-3 and ICE ($\theta_I = 0.29$). This means that, after the gene duplication, some amino acid sites may have been involved in the functional divergence

(a)

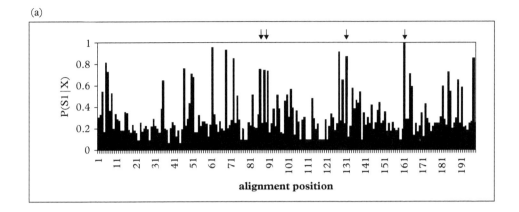

(b)

site		CED-3	ICE
161	Sequence conservation	An invariant Trp (W)	Highly variable
	Structural features	Form a narrow pocket with an extra loop; form a H-bond	No extra loop; a shallow depression found
	Substrate specificity	Network with a group of amino acids; Hydrophilic side chains	Hydrophobic side chains
86/88	Structural features	No surface loop	Lie in an extra surface loop
131	Sequence conservation	Highly variable	Highly conserved
	Structural features	Not a cleavage site	Cleavage site for proenzyme processing

Figure 4.3 *(A) Site-specific profile for predicting critical amino acid residues responsible for the (type-I) functional divergence between CED-3 and ICE subfamilies, measured by the posterior probability of functional divergence being related at each site. The arrows point to four amino acid residues at which functional divergence between CED-3 and ICE has been verified by experimentation; see (B) for details.*

between CED-3 and ICE. We further carried out the posterior profile analysis (Fig. 4.3) and predicted 29 crucial amino acid residues that are responsible for functional divergence between them at the cutoff of > 70% (posterior probability), which were mapped onto the 3-D structure of caspases. The resolved X-ray crystal structures of human caspase-1 and -3 (Wilson et al. 1994; Rotonda et al. 1996) were used to illustrate the structural features of ICE and CED-3 subfamilies, respectively (Fig. 4.4).

From the literature, Wang and Gu (2001) found some experimental evidence for four predicted residues that are involved in the functional-structural divergence between CED-3 and ICE subfamilies (Fig. 4.3 and Fig. 4.5):

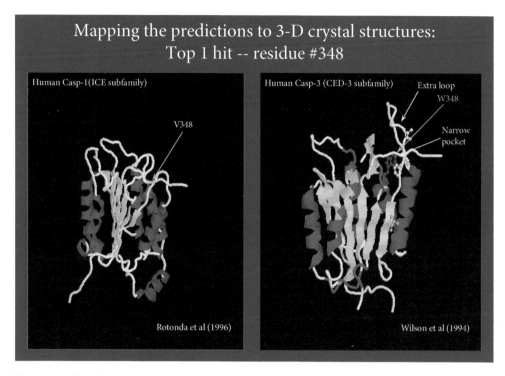

Figure 4.4 *Protein structures of human caspase-1 (ICE-type) and human caspase-3 (CED-3 type).*

1. Residue 161(348): (In the literature, this site is numbered as W348, according to the protein sequence of human caspase-1) is critical for CED-3 caspase substrate specificity by interacting with a unique surface loop in the 3-D structure $[P(F_1|X) = 0.999]$ (Rotonda et al. 1996). At this position, all 22 sequences from the CED-3 subfamily contain an invariant tryptophan (W), whereas a variety of residues are present in the ICE subfamily (Fig. 4.5). Crystal structural analysis (Fig. 4.4) reveals that W348 is a key determinant for the caspase-3 (CED-3)-type specificity. First, W348 forms a narrow pocket with the surface loop that is highly conserved in the CED-3 subfamily. The steric constriction due to this pocket determines the preference of caspase-3 for the substrates with small hydrophilic side chains. Second, W348 along with a group of residues forms a hydrogen bond network, which affects the interaction with the substrate. In contrast, the surface loop shared with CED-3 caspases seems to be deleted in all ICE-type caspases, as shown in the boxed region in the multiple alignment (Fig. 4.5). Hence, the relaxed evolutionary constraint observed at this position in the ICE subfamily is likely to be caused by the 3-D structural difference.

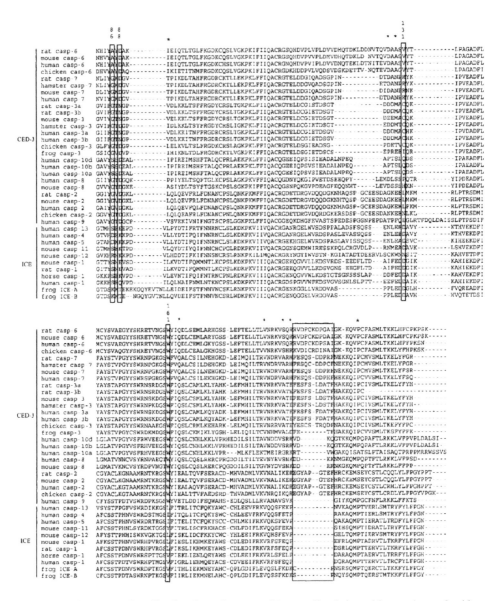

Figure 4.5 *Alignment of predicted regions of caspases. Four predicted sites with experimental evidence are highlighted. The sites with asterisks are predicted residues. The boxed region in the C-terminus is the critical region for CED-3 substrate specificity: Most CED-3 type caspases form a surface loop, whereas a shallow depression is found in ICE-type caspases.*
Modified from Wang and Gu (2001).

2. **Residues 86** [$P(F_1|X) = 0.75$] **and 88** [$P(F_1|X) = 0.74$]: They are responsible for 3-D difference with an unknown functional role. Indeed, in human caspase-1 (ICE), these two residues appear to lie in a small loop that is not found in the CED-3 subfamily.

3. **Residue 131** [$P(F_1|X) = 0.866$]: It is proteolytic site specific sites to the ICE subfamily. All caspases are synthesized as inactive proenzymes that need to be processed to the mature forms (Nicholson et al. 1995). However, distinct cleavage sites within the precursors are found for two subfamilies. D131 is known as a cleavage site in human caspase-1 ICE type. All ICE-type caspases preserve an aspartic acid (D) at this position, except for mouse caspase-12 (asparagine, E). However, human caspase-3 (CED-3 type) utilizes two other asparagine sites for cleavage (Rotonda et al. 1996) so that the functional role of position 131 in CED-3caspases is no longer important. Therefore, the altered evolutionary constraints at this position can be well explained by the different utilization of cleavage sites for the precursor processing between CED-3 and ICE subfamilies.

4.5 Type-II functional divergence between GPCR subfamilies GCGR and GLP-1R

4.5.1 Predicting type-II functional divergence residues

The glucagon-like family belongs to secretin type GPCRs, which constitutes four hormone receptors duplicated from the early stage of vertebrates. These receptors play crucial roles in hormonal homeostasis in humans and other animals, and serve as important drug targets for several endocrine disorders. To this end, Sa et al. (2017) analyzed type-II functional divergence between GCGR and GLP-1R of glucagon-like family to identify residues conserved in functional constraints but different in physicochemical properties. Based on the neighbor-joining tree(Fig. 4.6A), the coefficient of type-II functional divergence (θ_{II}) was estimated between GCGR and GLP-1R is 0.236 ± 0.052, which is significantly higher than 0 ($P < 0.001$).

Rejection of the null hypothesis $\theta_{II} = 0$ indicates that after gene duplication, some amino acid residues that were conserved in both GCGR and GLP-1R clusters but had radically changed their amino acid properties between them. Sa et al. (2017) used the posterior probability Q_{II} to identify amino acid residues critical in type-II functional divergence between them (Fig. 4.6B). Using an empirical cutoff of $Q_{II} > 0.67$ (the posterior ratio $R_{II} > 2$), the authors predicted eight type-II functional divergence-related residues between paralogous GCGR and GLP-1R (Fig. 4.6C, D).

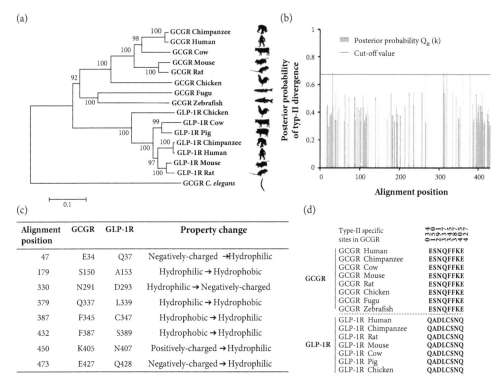

Figure 4.6 *Analytical pipeline for type-II functional divergence between GCGR and GLP-1R. (A) Phylogenetic tree of GCGR and GLP-1R. (B) Site-speci?c pro?le for predicting critical amino acid residues responsible for type-II functional divergence between GCGR and GLP-1R measured by posterior probability QII (k). (C) Overview of amino acid changes in the eight predicted sites in type-II functional divergence. (D) Sequence conservation analysis of the two clusters for GCGR and GLP-1R. GCGR, glucagon receptor; GLP-1R, glucagon-like peptide 1 receptor.*

Reproduced with permission from Sa, Zhou, Zou, Su, Gu, 'Paralog-divergent Features May Help Reduce Off-target Effects of Drugs: Hints from Glucagon Subfamily Analysis', *Genomics, Proteomics & Bioinformatics*, 15, (4), 2017, Pages 246–254, Elsevier

4.5.2 Type-II functionally-divergent residues in binding sites of anti-diabetic drugs

The issues of cross-reactivity arising from paralogs have been well recognized. Identifying paralog-divergent features as targetable differences might be helpful in paralog discrimination and has already been implemented in therapeutic drug design. The GCGR antagonist MK-0893 is used to treat patients with type 2 diabetes to substantially reduce fasting and postprandial glucose concentrations. MK-0893 acts at allosteric binding sites of the seven transmembrane helical domain (7TM) in positions among TM5, TM6, and TM7 in GCGR (Fig. 4.7A).

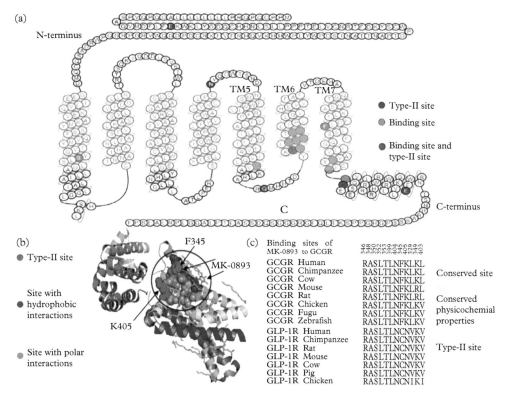

Figure 4.7 *Paralog-divergent features are considered targetable differences of drugs. (A) Snake-plot diagram of GCGR with annotation of important residues. (B) Different physicochemical properties of bipartite antagonist pocket corresponding to the dual polar/hydrophobic binding cleft in GCGR. (C) Sequence conservation analysis of 12 binding sites of MK-0893 to GCGR.*

Reproduced with permission from Sa, Zhou, Zou, Su, Gu, 'Paralog-divergent Features May Help Reduce Off-target Effects of Drugs: Hints from Glucagon Subfamily Analysis', *Genomics, Proteomics & Bioinformatics*, 15, (4), 2017, Pages 246–254, Elsevier

It was found that two predicted sites of type-II functional divergence between GCGR and GLP-1R, F345 and K405, were in the binding sites of MK-0893 to GCGR. F345 was hydrophobic in GCGR but its homologous site in GLP-1R is hydrophilic. If a molecule of drug is designed to be hydrophobic, it tends to bind to the hydrophobic F345 in GCGR rather than the hydrophilic residue in GLP-1R. Another type-II specific site K405 was positively-charged in GCGR while its homologous site in GLP-1R was electrically neutral. Thus a molecule of drug designed to be negatively-charged are more likely to interact with positively-charged K405 in GCGR instead of binding to the electrically neutral residue in GLP-1R. Because the physio-chemical properties of amino acids play an important role in the interaction of protein receptors with their ligands (small molecules, peptides, agonists, and antagonists), changes in their physicochemical nature and conformation may reduce cross-reactivity due to the binding of antagonist drugs to unexpected paralogs. Therefore, determining type-II functional divergence-related sites between two paralogs is effective for identifying targetable differences in therapeutic drug design.

References

Casari, G., Sander, C., and Valencia, A. (1995). A method to predict functional residues in proteins. *Nature Structural Biology* 2, 171–178.

Dayhoff, M.O. (1978). Protein segment dictionary 78. (National Biomedical Research Foundation).

Dermitzakis, E.T., and Clark, A.G. (2001). Differential selection after duplication in mammalian developmental genes. *Molecular Biology and Evolution* 18, 557–562.

Fitch, W.M. (1971). Toward defining the course of evolution: Minimum change for a specific tree topology. *Systematic Biology*, 20, 406–416.

Gu, J., Wang, Y., and Gu, X. (2002). Evolutionary analysis for functional divergence of Jak protein kinase domains and tissue-specific genes. *Journal of Molecular Evolution* 54, 725–733.

Gu, X. (1999). Statistical methods for testing functional divergence after gene duplication. *Molecular Biology and Evolution* 16, 1664–1674.

Gu, X. (2001a). Mathematical modeling for functional divergence after gene duplication. *Journal Computational Biology* 8, 221–234.

Gu, X. (2001b). Maximum-likelihood approach for gene family evolution under functional divergence. *Molecular Biology and Evolution* 18, 453–464.

Gu, X., Wang, Y., and Gu, J. (2002). Age distribution of human gene families shows significant roles of both large- and small-scale duplications in vertebrate evolution. *Nature Genetics* 31, 205–209.

Gu, X. (2006). A simple statistical method for estimating type-II (cluster-specific) functional divergence of protein sequences. *Molecular Biology and Evolution* 23, 1937–1945.

Gu, X., Fu, Y.-X., and Li, W.-H. (1995). Maximum likelihood estimation of the heterogeneity of substitution rate among nucleotide sites. *Molecular Biology and Evolution* 12, 546–557.

Gu, X., and Nei, M. (1999). Locus specificity of polymorphic alleles and evolution by a birth-and-death process in mammalian MHC genes. *Molecular Biology and Evolution* 16, 147–156.

Gu, X., and Zhang, J. (1997). A simple method for estimating the parameter of substitution rate variation among sites. *Molecular Biology and Evolution* 14, 1106–1113.

Holland, P.W., Garcia-Fernandez, J., Williams, N.A., and Sidow, A. (1994). Gene duplications and the origins of vertebrate development. *Development* 1994, 125–133.

Jones, D.T., Taylor, W.R., and Thornton, J.M. (1992). The rapid generation of mutation data matrices from protein sequences. *Computer Applications in the Biosciences* 8, 275–282.

Kimura, M. (1983). Rare variant alleles in the light of the neutral theory. *Molecular Biology and Evolution* 1, 84–93.

Landgraf, R., Xenarios, I., and Eisenberg, D. (2001). Three-dimensional cluster analysis identifies interfaces and functional residue clusters in proteins. *Journal of Molecular Biology* 307, 1487–1502.

Lichtarge, O., Bourne, H.R., and Cohen, F.E. (1996). An evolutionary trace method defines binding surfaces common to protein families. *Journal of Molecular Biology* 257, 342–358.

Livingstone, C.D., and Barton, G.J. (1996). Identification of functional residues and secondary structure from protein multiple sequence alignment. *Methods Enzymol* 266, 497–512.

Nicholson, D.W., Ali, A., Thornberry, N.A., et al. (1995). Identification and inhibition of the ICE/CED-3 protease necessary for mammalian apoptosis. *Nature* 376, 37–43.

Ohno, S. (1970). Evolution by gene duplication (Springer Berlin).

Rotonda, J., Nicholson, D.W., Fazil, K.M., et al. (1996). The three-dimensional structure of apopain/CPP32, a key mediator of apoptosis. *Nature Structural & Molecular Biology* 3, 619–625.

Sa, Z., Zhou, J., Zou, Y., et al. (2017). Paralog-divergent Features May Help Reduce Off-target Effects of Drugs: Hints from Glucagon Subfamily Analysis. *Genomics Proteomics and Bioinformatics* 15, 246–254.

Su, Z., Wang, J., Yu, J., Huang, X., and Gu, X. (2006). Evolution of alternative splicing after gene duplication. *Genome Research* 16, 182–189.

Wakeley, J. (1993). Substitution rate variation among sites in hypervariable region 1 of human mitochondrial DNA. *Journal of Molecular Evolution* 37, 613–623.

Wang, Y., and Gu, X. (2001). Functional divergence in the caspase gene family and altered functional constraints: statistical analysis and prediction. *Genetics* 158, 1311–1320.

Wilson, K.P., Black, J.A., Thomson, J.A., et al. (1994). Structure and mechanism of interleukin-1 beta converting enzyme. *Nature* 370, 270–275.

Wolfe, K.H., and Shields, D.C. (1997). Molecular evidence for an ancient duplication of the entire yeast genome. *Nature* 387, 708–713.

Yang, Z. (1993). Maximum-likelihood estimation of phylogeny from DNA sequences when substitution rates differ over sites. *Molecular Biology and Evolution* 10, 1396–1401.

Yang, Z., Kumar, S., and Nei, M. (1995). A new method of inference of ancestral nucleotide and amino acid sequences. *Genetics* 141, 1641–1650.

Zheng, Y., Xu, D., and Gu, X. (2007). Functional divergence after gene duplication and sequence-structure relationship: a case study of G-protein alpha subunits. *Journal of Experimental Zoology Part B, Molecular and Developmental Evolution* 308, 85–96.

5

Rank of Genotype-Phenotype Map and Effective Gene Pleiotropy Estimation

Like many biological terminologies, the concept of pleiotropy is easy to understand but difficult to measure (Chevin et al. 2010; Cooper et al. 2007; Dudley et al. 2005; Martin and Lenormand 2006; MacLean et al. 2004; Razeto-Barry et al. 2011, 2012; Gu 2014; Zhang and Hill 2003; Wingreen et al. 2003; Welch and Waxman 2003). Practically, biologists address this issue by experimentally identifying distinct phenotypes, *in vitro* or *in vivo*, caused by the same mutation (mutational pleiotropy) or mutations from the same gene (gene pleiotropy). Although these studies largely focused on the nature of phenotypes, the results did provide a minimum estimate about the extent of pleiotropy. Thus, one may compare whether a gene is more pleiotropic than others, as long as the same technological platform is used, e.g., mouse knockout experiments.

However, different pleiotropy measures based on different technological platforms certainly represent biologically significant differences that may even lead to contradictory interpretations (Wagner and Zhang 2011; Paaby and Rockman 2013; Gu 2014). To illustrate this multi-faceted problem of phenotypic pleiotropy, we invoke the basic system formulated by Paaby and Rockman (2013), which has three components:

1. Molecular pleiotropy (f) refers to the number of protein functions, most likely *in vitro*. For instance, an enzyme is called pleiotropic if it can catalyze many distinct substrates.

2. Developmental pleiotropy (d) refers to the number of developmental pathway autonomies that underlie diverse phenotypes, including diseases. Developmental pleiotropy provides concrete examples to demonstrate how a single gene or mutation may affect distinct phenotypes *in vivo*.

3. Selectional pleiotropy (n) refers to the number of molecular phenotypes (Gu 2007), each of which corresponds to a single nontrivial fitness component. In

Statistical Analysis of Molecular and Genomic Evolution. Xun Gu, Oxford University Press. © Xun Gu (2024).
DOI: 10.1093/oso/9780198816515.003.0005

essence, selectional pleiotropy is the dimension of Fisher's geometric model, which underlies most statistical studies of the genotype-phenotype map (Fisher 1930; Lande 1980; Turelli 1985; Wagner 1989; Barton 1990).

5.1   Rank of genotype-phenotype map

5.1.1   Geometric construction of genotype-phenotype map

Figure 5.1 schematically presents the map from the genotype to the phenotype. Intuitively, it can be viewed as a three-layer structure as follows.

Genotype dimensionality: The complexity of a given genotype resource (G) can be quantified as the dimension of mutational effects (r), which is the canonical mutation types of each genotype unit. If single mutations are treated as genotypes, for instance,

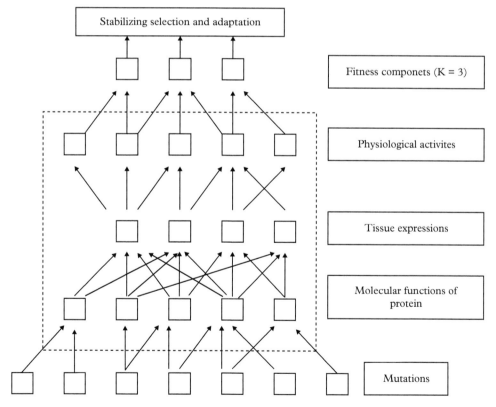

Figure 5.1 *Schematic presentation for the concept of molecular phenotypes that affect the fitness of an organism.*
Figure from Gu (2007).

we have $r = 1$, which means that, though it can affect multiple aspects of phenotypes (mutational pleiotropy), there are no degrees of freedom simply because the same biochemical-structural property caused by this single mutation underlies most of the phenotypic consequences. If substitutions at nucleotide sites are treated as genotypes, we have $1 \leq r \leq 3$ because potentially one nucleotide site has three different substitutions, but a genotype sample may have no nucleotide sites with three substitutions, resulting in a dimension degeneration.

Phenotype dimensionality: From the evolutionary view, the phenotype space includes an N-layer biological system, which ultimately points to the organismal fitness with n components. Let P_i, $i = 1, \ldots, N$ be the dimension of the i-th phenotype layer. While this N-layer biological system is abstract, the lowest level is the molecular functions ($P_1 = f$) related to activities of gene product. Let $P_{\min}$ be the minimum dimension among N biological layers, i.e., $P_{min} = \min(P_1, \ldots, P_N)$.

Minimum pleiotropy   The dimension of each layer (P_i, $i = 1, \ldots, N$) can be considered as a measure of pleiotropy, a well-known multi-faceted problem of pleiotropy that could be empirically estimated under technical platforms but the magnitudes usually cannot be compared to each other. On the other hand, pleiotropy can be theoretically defined as the related number of fitness components (n). To clarify these confusions, we propose the concept of minimum pleiotropy as

$$n = P_{\min} \tag{5.1}$$

That is, the number of fitness components is the same as the minimum dimension of the multi-layer biological systems. While it may provide one solution to the many-face problem, the challenging problem is under what condition we are able to estimate $n = P_{\min}$. To this end, we should address the rank of the genotype-phenotype map.

5.1.2   Rank of genotype-phenotype map

Without loss of generality, the genotype-phenotype map can be schematically represented as $G \rightarrow P_1 (\text{or } f) \rightarrow (P_2, \ldots, P_{N-1}) \rightarrow n(\text{fitness})$, where $(P_2, \ldots, P_{N-1})$ represents the black box of phenotype complexity. We then define the rank of the genotype-phenotype map (K) as the canonical number of mapping paths from the genotype (G) to the fitness components (n) through the entire phenotypic space. In other words, K is the degrees of freedom for a particular genotype-phenotype mapping, which highly depends on the nature of the genotype. Putting this together, the rank (K) of the genotype-phenotype map is the smaller one of the rank of phenotypic complexity (minimum phenotypic pleiotropy) and the rank of mutational effects; that is,

$$K = \min(r, P_{\min}) = \min(r, n) \tag{5.2}$$

(Gu 2014). Intuitively, r is the number of starting points of map paths; obviously, $r = 1$ for single mutations and so $K = 1$.

5.2 Statistical Formulation of $K = (r, n)$

In this section we construct the genotype-phenotype map from the theory of statistical genetics (Wagner 1989) and the stabilizing selection model of protein sequence evolution (Gu 2007), with a constraint of the rank of genotype-phenotype map.

5.2.1 Rank of phenotype-genotype matrix (A)

Suppose that a gene can affect n molecular phenotypes, denoted by $y_1, \ldots, y_n$, respectively. Under the Gaussian-like stabilizing selection model, the fitness function of $\mathbf{y} = (y_1, \ldots, y_n)'$ is given by

$$w(\mathbf{y}) = \exp \left\{ -\frac{\mathbf{y}' \Sigma_w^{-1} \mathbf{y}}{2} \right\} \tag{5.3}$$

where Σ_w is the (positive-definite) matrix of selection strengths. It is known that the coefficient of selection for any mutation that results in a deviation of molecular phenotypes ($\mathbf{y}$) from the optimal values ($\mathbf{0}$) is approximately given by $s(\mathbf{y}) = w(\mathbf{y}) - 1 \approx -\mathbf{y}' \Sigma_w^{-1} \mathbf{y}/2$. It follows that the selection intensity ($S = 4N_e s$) is given by

$$S(\mathbf{y}) = -2N_e \left(\mathbf{y}' \Sigma_w^{-1} \mathbf{y} \right) \tag{5.4}$$

On the other hand, we assume that the effects on molecular phenotypes of a mutation follows a multi-normal distribution with the covariance matrix Σ_m, that is,

$$p(\mathbf{y}) = N(\mathbf{y}; 0, \Sigma_m) \tag{5.5}$$

The genotype-phenotype map is then statistically represented by the S-distribution of mutations, $\Phi(S)$, characterized by the genotype-phenotype matrix

$$\mathbf{A} = \Sigma_w^{-1} \Sigma_m \tag{5.6}$$

Therefore, the rank of genotype-phenotype map (K) is defined as the rank of matrix $\mathbf{A}$. Let $\alpha_1, \ldots, \alpha_n$ be the eigenvalues of $\mathbf{A}$. It appears that the rank (K) of matrix $\mathbf{A}$ equals the number of non-zero eigenvalues, which determines the true dimensionality of the genotype-phenotype map.

5.2.2   Construction of the genotype-phenotype matrix (A)

A two-step map

Without loss of generality, we assume that the rank of Σ_w (selection) is equal to the number (n) of molecular phenotypes, but the rank of Σ_m (mutation) could be less than n. We use the idea of Wagner (1989) to demonstrate this issue. Let x_i be the genotype value of the i-th functional unit ($i = 1, \ldots, r$). To establish the relationship between molecular functions $\mathbf{x} = (x_1, \ldots, x_r)'$ and molecular phenotypes $\mathbf{y} = (y_1, \ldots, y_n)'$, we define an $n \times r$ matrix ($\mathbf{Q}$) such that

$$\mathbf{y} = \mathbf{Q}\mathbf{x} \tag{5.7}$$

where one may restrict $0 \leq q_{ij} \leq 1$. The coefficients q_{ij} of the "physiological matrix" $\mathbf{Q}$ represent the influence of the j-th functional unit (x_j) on the i-th molecular phenotype y_i. The j-th column vector ($q_{1j}, \ldots, q_{nj}$) represents the spectrum of all pleiotropic effects of the j-th functional unit, while each row vector ($q_{i1}, \ldots, q_{ir}$) represents the mutational influence of all r functional units on a given molecular phenotype y_i. For simplicity we assume that a single mutation affects a single functional unit, and the mutational effect on the i-th functional unit follows a normal distribution with mean zero and the variance v_i^2. Then, one can show that the mutational covariance matrix is given by

$$\Sigma_m = \mathbf{Q}\,\text{diag}(v_1^2, \ldots, v_r^2)\mathbf{Q}' \tag{5.8}$$

According to the matrix theory, we then claim that when the number of functional units is equal to or greater than the number of molecular phenotypes ($r \geq n$), the rank of Σ_m is n (full-rank). Hence, the rank of matrix $\mathbf{A} = \Sigma_w^{-1}\Sigma_m$ is n, and all its eigenvalues are positive. By contrast, when $r < n$, the rank of Σ_m is r so that the rank of matrix $\mathbf{A}$ is r that is less than n. In this case, there are r positive eigenvalues and $n - r$ zero eigenvalues of $\mathbf{A}$ in Σ_m. Putting together, we show that gene pleiotropy defined by $K = (n, r)$ is the rank of the genotype-phenotype matrix $\mathbf{A}$.

A multi-step map

Consider the case where the genotype-phenotype map is schematically represented as a linear mapping model $G \to f \to (P_2, \ldots, P_{N-1}) \to n$, which can be characterized by three matrices (Wagner 1989): an $r \times r$ variance-covariance matrix ($\mathbf{V}$) for the correlated mutational effects of genotypes (G), an $r \times n$ transformation matrix $\mathbf{T}$ that links between r, the number of mutational effects, and n, the number of fitness components (molecular phenotypes), and an $n \times n$ stabilizing selection matrix Σ_w that is the full rank of n. The linear mapping model claims that the mutation matrix V and transformation matrix T together determine how mutations ultimately affect n molecular phenotypes, as characterized by an $n \times n$ matrix defined by

$$\Sigma_m = T'VT \tag{5.9}$$

However, the rank of Σ_m is uncertain because T is a multiplication of those consecutive step-transformation matrices; that is, $T = T_1, T_2, \ldots, T_N$, where T_1 is the matrix with r rows and P_1 columns, T_2 is the one with P_1 rows and P_2 columns, and so forth. Thus, linear algebra theory states that the rank of Σ_m is the minimum of $(r, P1, \ldots, P_{N-1}, P_N)$. Noting that for the minimum pleiotropy $P_{min} = \min(P_1, \ldots, P_{N-1}, P_N)$, we have the rank of Σ_m given by $\min(r, P_{min})$.

Under this transformation model, the rank (K) of the genotype-phenotype map turns out to be the rank of matrix $A = \Sigma_w^{-1}\Sigma_m$. As Σ_w is a full rank of n, the rank of A is the same as that of Σ_m. Thus, we have shown $K = \min(r, P_{min}) = \min(r, n)$. The equals sign holds when $n = P_N$ is the smallest among $P_1, \ldots, P_N$ (Chen et al. 2013).

5.3 Rank of genotype-phenotype map: estimation

5.3.1 Principle of Gu's 2007 method

Evolutionary rate of protein sequence

According to the theory of molecular evolution (Kimura 1983), the evolutionary rate (λ) of a mutant affecting the molecular phenotypes $(\mathbf{y})$ can be written as

$$\lambda(\mathbf{y}) = v\frac{S(\mathbf{y})}{1 - e^{-S(\mathbf{y})}} \tag{5.10}$$

where v is the mutation rate. Thus, given $p(\mathbf{y})$, the distribution of mutational effects that generates a variation of $\mathbf{y}$, the k-th moment of evolutionary rate of a gene is therefore given by

$$\bar{\lambda}^k = v^k \int \left[\frac{S(\mathbf{y})}{1 - e^{-S(\mathbf{y})}} \right]^k p(\mathbf{y}) d\mathbf{y} \tag{5.11}$$

The mean (first moment) of evolutionary rate

Under the stabilizing selection model that assures $S(\mathbf{y}) < 0$, one can approximately solve the problem of any k-th moment of evolutionary rate $(\bar{\lambda}^k)$. To this end, Gu (2007)

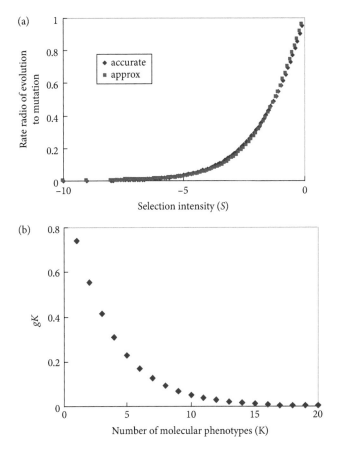

Figure 5.2 *(A) The λ/v-S (the rate ratio of evolution to mutation versus the selection intensity) relationship, for the accurate formula and the approximation. (B) The g_K-K plotting for estimating the effective number of molecular phenotypes.*
Figure from Gu (2007).

used the approximation $S/(1 - e^{-S}) \approx e^{-|S|}(1 + c|S|)$, which holds well for $S < 0$, where $c \approx 0.5772$ (Fig. 5.2). First we consider the mean evolutionary rate $\bar{\lambda}$ as given by

$$\bar{\lambda} = v \int e^{-2N_e(\mathbf{y}'\boldsymbol{\Sigma}_w^{-1}\mathbf{y})} \left[1 + c(2N_e)(\mathbf{y}'\boldsymbol{\Sigma}_w^{-1}\mathbf{y})\right] p(\mathbf{y}) d\mathbf{y} \tag{5.12}$$

It has been shown that the analytical solution of $\bar{\lambda}$ depends on the rank of genotype-phenotype matrix $\mathbf{A} = \boldsymbol{\Sigma}_w^{-1}\boldsymbol{\Sigma}_m$. Let $\alpha_1, \alpha_2, \ldots, \alpha_n$ be the (nonnegative) eigenvalues of $\mathbf{A}$ (in descending order).

$$\bar{\lambda} = v \prod_{i=1}^{K} [1 + 2B_i]^{-1/2} \left(1 + c \sum_{i=1}^{K} \frac{B_i}{1 + 2B_i}\right) \tag{5.13}$$

where $B_i = 2N_e\alpha_i$. That is, the evolutionary rate of a gene is determined by the mutation rate (v), gene pleiotropy (K) measured by the number of molecular phenotypes, and a number of baseline selection intensities (B_i) of molecular phenotypes. Note that under the SM_w condition, all $B_i \geq 0$.

The second-moment of evolutionary rate

In addition to the mean evolutionary rate in Eq. (5.10), we consider the second-moment of the evolutionary rate, $\bar{\lambda}^2$. From the general formula

$$\bar{\lambda}^2 = v^2 \int \left[\frac{S(\mathbf{y})}{1 - e^{-S(\mathbf{y})}} \right]^2 p(\mathbf{y})d\mathbf{y} \tag{5.14}$$

With a close approximation, Gu (2007) has derived the second moment of λ as follows

$$\bar{\lambda}^2 = v^2 \prod_{i=1}^{K} [1 + 4B_i]^{-1/2} \left[\left(1 + c \sum_{i=1}^{K} \frac{B_i}{1 + 4B_i} \right)^2 + c^2 \sum_{i=1}^{K} \frac{2B_i^2}{(1 + 4B_i)^2} \right] \tag{5.15}$$

where $c \approx 0.5772$.

5.3.2 Effective gene pleiotropy (K_e)

The mean $(\bar{\lambda})$ and the second moment $(\bar{\lambda}^2)$ of the evolutionary rate suggest that, K, the rank of genotype-phenotype map, or the nontrivial number of distinct B_i's, can be estimated if one can calculate $\bar{\lambda}$ and $\bar{\lambda}^2$ from the sequence data. However, the difficulty is how to evaluate B_i's. Gu (2007) developed an effective approach to solving this problem: extensive numerical and simulation analyses showed that $\bar{\lambda}$ and $\bar{\lambda}^2$ can be well approximated as follows

$$\frac{\bar{\lambda}}{v} \approx \prod_{i=1}^{K} [1 + 2B_i]^{-1/2} \left(1 + cK/2 \right)$$

$$\frac{\bar{\lambda}^2}{v^2} \approx \prod_{i=1}^{K} [1 + 4B_i]^{-1/2} \left(1 + cK/2 + c^2K/8 + c^2K^2/16 \right) \tag{5.16}$$

respectively. Next we calculate the ratio of second-moment to the mean of the evolutionary rate, normalized by the mutation rate (v), i.e., $g_K = [\bar{\lambda}^2/v^2]/[\bar{\lambda}/v]$, resulting in

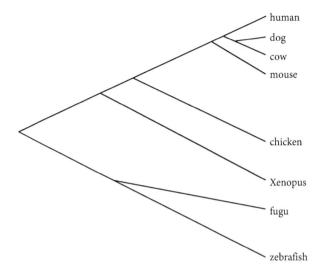

Figure 5.3 *The phylogenetic tree of vertebrates used in our data analysis.*

$$g_K = 2^{-K/2} \left[\prod_{i=1}^{K} \left(1 + \frac{1}{1+4B_i} \right)^{1/2} \right] \left[1 + \left(\frac{c}{4} \right)^2 \frac{K(K+2)}{1+cK/2} \right] \qquad (5.17)$$

We then define the effective gene pleiotropy (K_e) as the effective number of molecular phenotypes that have experienced strong stabilizing selection, i.e., with a large baseline selection intensity. Hence, K_e is less than the true number of molecular phenotypes. As for each i, the approximation $1 + 1/(1 + 4B_i) \approx 1$, we have approximately

$$g_{K_e} = 2^{-K_e/2} \left[1 + \left(\frac{c}{4} \right)^2 \frac{K_e(K_e+2)}{1+cK_e/2} \right] \qquad (5.18)$$

which is only K_e-dependent ($c \approx 0.5772$). As shown in Fig. 5.3, g_{K_e} decreases when K_e increases; $g_{K_e} = 1$ when $K_e = 0$, and $g_{K_e} \to 0$ when $K_e \to \infty$.

5.3.3   Estimation of K_e

Equation (5.18) indicates that the effective gene pleiotropy (K_e) can be estimated if the g_K-measure can be estimated from the protein sequence. To this end, we use a common measure for $\bar{\lambda}/v$, that is, the ratio of nonsynonymous to synonymous substitutions (d_N/d_S). The second moment of rate $\bar{\lambda^2}/v^2$ is related to the H-measure (Gu et al. 1995) for the rate variation among sites, defined by

$$H = 1 - (\bar{\lambda})^2/\bar{\lambda^2} \qquad (5.19)$$

Ranging from 0 to 1, a higher value of H means a greater rate variation among sites, and *vice versa*. Therefore, by relating the first and second moments of the evolutionary rate to d_N/d_S and H, respectively, one can show that

$$g_K = \frac{d_N/d_S}{1-H} = 2^{-K_e/2}\left[1 + \left(\frac{c}{4}\right)^2 \frac{K_e(K_e+2)}{1+cK_e/2}\right] \tag{5.20}$$

Moreover, numerical analysis shows that the above equation can be well-approximated by

$$\frac{d_N/d_S}{1-H} \approx e^{-0.301K_e} \tag{5.21}$$

Gu (2007) implemented a computational procedure to estimate K_e, which included the following steps:

1. Infer the phylogenetic tree from a multiple alignment of homologous protein sequences.

2. Estimate the nonsynonymous to synonymous ratio (d_N/d_S) from closely-related species.

3. Use the method of Gu and Zhang (1997) to infer the (bias-corrected) number of changes at each site, under the inferred phylogeny. Let $\bar{x}$ and $V(x)$ be the mean and variance of number of changes over sites, respectively. Assuming a Poisson process at each site, we obtain the mean evolutionary $\bar{\lambda} = \bar{x}/T$, where T is the total evolutionary time along the tree. Similarly, the variance of evolutionary rate among sites is given by $V(\lambda) = [V(x) - \bar{x}]/T^2$.

4. It follows that H can be estimated by

$$\hat{H} = \frac{V(x) - \bar{x}}{V(x) + \bar{x}(\bar{x} - 1)} \tag{5.22}$$

5. Estimate K_e by solving Eq. (6.20) numerically, or approximating based on Eq. (6.21) by

$$K_e \approx -3.322 \ln\left[\frac{d_N/d_S}{1-H}\right] \tag{5.23}$$

6. The sampling variance of K_e can be approximately calculated by the delta-method, that is,

$$Var(K_e) \approx 3.322^2 \left[\frac{V(d_N/d_S)}{(d_N/d_S)^2} + \frac{V(H)}{(1-H)^2}\right] \tag{5.24}$$

While the variance of d_N/d_S can be estimated conventionally, we implement a bootstrapping approach to calculate the variation of H, $V(H)$.

5.3.4   Estimation of effective selection intensity

In addition to K_e, there are two important evolutionary parameters: the mean selection intensity $\bar{S}$ and the baseline selection intensity (B_0). Their relationship can be written by $\bar{S} = -K \times B_0$. Replacing $\bar{\lambda}/v$ by d_N/d_S in Eq. (5.15), we have

$$\frac{d_N}{d_S} = \prod_{i=1}^{K}(1 + 2B_i)^{-1/2}\left(\left(1 + \frac{cK}{2}\right)\right) = \left(1 + \tilde{B}_0\right)^K\left(\left(1 + \frac{cK}{2}\right)\right) \tag{5.25}$$

where $\tilde{B}_0$ is the effective baseline selection intensity defined by

$$\left(1 + \tilde{B}_0\right)^K = \prod_{i=1}^{K}(1 + 2B_i) \tag{5.26}$$

After replacing K by K_e, one can estimate $\tilde{B}_0$ as follows

$$\tilde{B}_0 = \frac{1}{2}\left\{\left[\frac{1 + cK_e/2}{d_N/d_S}\right]^{2/K_e} - 1\right\} \tag{5.27}$$

as well as the mean selection intensity by

$$\bar{S} = -\frac{K_e}{2}\left\{\left[\frac{1 + cK_e/2}{d_N/d_S}\right]^{2/K_e} - 1\right\} \tag{5.28}$$

5.3.5   Bias-corrected estimation of effective gene pleiotropy

Though K_e provides a conserved estimate for the degree of gene pleiotropy, it is desirable to correct the underestimation bias without introducing additional assumptions. To this end, we consider the following problem: When the function g is given, what is the difference between K_e and the true K? To be concise, let $\phi(K) = (c^2/4)K(K+2)/(1 + cK/2)$ and so $\phi(K_e)$. By equating Eq. (5.16) with Eq. (5.17), we have

$$2^{-K/2}\left[\prod_{i=1}^{K}\left(1 + \frac{1}{1 + 4B_i}\right)^{1/2}\right][1 + \phi(K)] = 2^{-K_e/2}[1 + \phi(K_e)] \tag{5.29}$$

After some simple algebra, it turns out to be the following relationship between K_e and the true pleiotropy K

$$K_e = K - \sum_{i=1}^{K} \log_2(1 + \frac{1}{1+4B_i}) + 2\log_2 \frac{1 + \phi(K)}{1 + \phi(K_e)}$$

$$\approx K - \sum_{i=1}^{K} \log_2(1 + \frac{1}{1+4B_i}) \tag{5.30}$$

where the last approximation holds because of $\phi(K) \approx \phi(K_e)$ so that the third term on the right hand is usually negligible.

Then, we propose a bias-corrected estimate of effective gene pleiotropy ($\tilde{K}$) by replacing B_i's with the estimated baseline selection intensity $\tilde{B}_0$, leading to

$$\tilde{K} = \frac{K_e}{1 - \eta} \tag{5.31}$$

where η is given by

$$\eta = \log_2(1 + \frac{1}{1 + 4\tilde{B}_0}) \tag{5.32}$$

5.4 From effective gene pleiotropy to fitness pleiotropy

5.4.1 Conditions for K_e interpreted as rank of genotype-phenotype map

Stabilizing selection with weak micro-adaptation

Under the stabilizing model (Turelli 1985; Waxman and Peck 1998), the population mean of molecular phenotypes is always fixed at the optimum (μ). Here we consider a general stabilizing selection model when the fitness optimum (μ) is no longer fixed during evolution. Rather, μ can be shifted by either environmental changes or internal physiological perturbations (Hartl and Taubes 1996; 1998; Poon and Otto 2000). Each shift of μ results in a micro-adaptation toward new fitness optimum, as called the *S*tabilizing selection with *M*icro-adaptation (*SM*) model.

Gu (2007) has shown that under the *SM* model, the coefficient of selection of **y** is given by $\rho(\mathbf{y}) \approx -\mathbf{y}'\mathbf{U}^{-1}\mathbf{y}/2$, where the matrix **U** is given by

$$\mathbf{U} = [\mathbf{\Sigma}_w^{-1} - \mathbf{\Sigma}_w^{-1}\mathbf{\Sigma}_\mu\mathbf{\Sigma}_w^{-1}]^{-1} \tag{5.33}$$

which characterizes the correlated nature of molecular phenotypes in fitness after stabilizing selection and micro-adaptation. It has been shown that the genotype-phenotype matrix $\mathbf{A}$ is given by

$$\mathbf{A} = \mathbf{U}^{-1}\mathbf{\Sigma}_m \tag{5.34}$$

Apparently, the stabilizing selection model is a special case when $\mathbf{\Sigma}_\mu = 0$ so that $\mathbf{U} = \mathbf{\Sigma}_w$.

5.4.2   Condition for K_e as a fitness pleiotropy

Universal rank assumption over sites

In this chapter, gene pleiotropy is specified by the means of selection pleiotropy, i.e., the distinct number (n) of fitness components affected by random mutations of the gene. Without loss of generality, n is defined by the minimum phenotype pleiotropy in the chain of the genotype-phenotype map. In other words, n is the minimum-required number of fitness components that can be affected by the gene function. Given the dimensionality (r) of the genotype, the rank of genotype-phenotype map is therefore simply given by

$$K = \min(r, n) \tag{5.35}$$

As introduced in above sections, Gu (2007) developed a statistical method to estimate K from a multiple sequence alignment (MSA) of orthologous genes from different species, without knowing the details of phenotype structure. Since the rank of genotype-phenotype map contains some invaluable information about the degree of gene pleiotropy, the terminology "effective gene pleiotropy" was introduced by Gu (2007); indeed, $K = n$ if $n < r$, and $K = r$ when $n > r$. While it is legitimate to call K the effective gene pleiotropy, Gu (2014) argued that over-use of this interpretation could be highly misleading, especially when r is small, such as in the study of Martin and Lenormand (2006), where $r = 1$, leading to $K = 1$ always, regardless of how large n is. Hence a fundamental issue is how we can choose appropriate sequence data such that the estimated rank of genotype-phenotype map (K_e) can be reasonably interpreted as an effective measure of gene pleiotropy.

The maximum genotype dimensionality (r_{max})

Intuitively, the uplimit of the genotype dimensionality for any given genotype (nucleotide) data is determined by the number of generic states (M). For instance, $M = 2$ for random nucleotides, $M = 4$ for aligned nucleotide sites, and $M = 20$ for aligned amino acid sites. In fact, the maximum rank (r_{max}) for a given genotype is the number of generic states minus one (the wild type), that is,

$$r_{max} = M - 1 \tag{5.36}$$

Principle of microergodicity

It should be noticed that the maximum genotype dimensionality (r_{max}) may not be reached for a given dataset. For instance, the genotype dimensionality of an amino acid site (r) could be much less than $r_{max} = 19$ if only a few mutations may have occurred; in other words, most genetic states are actually null. Theoretically, one may invoke the principle of "microergodicity" to address this issue. It claims that the mutational process can reach over all generic states for a sufficiently long time so that in this case we have $r \rightarrow r_{max}$. If the underlying sampling process has not had sufficient time for mutations to reach all states, $r << r_{max}$. However, due to purifying selections on deleterious mutations, only a small number of acceptable mutations could be observed even when the condition of microergodicity is satisfied. Hence, for a site with two amino acid types, one cannot decide whether it is due to the lack of evolutionary time.

We may use an empirical criterion to assess the condition of microergodicity. Suppose the mutation process at a site follows a Poisson process, with a rate of μ. Along a phylogeny with a total evolutionary time (T), the expected number of mutations is given by μT. Under the assumption that recurrent mutation is rare, a simple criterion is $\mu T > M - 1$. One may use synonymous mutation as a proxy of random neutral mutations. Let X_S be the expected number of synonymous substitutions per site along the phylogeny. In practice one may use the criterion

$$X_S > M - 1 \tag{5.37}$$

Analysis design

We also suggest, to be practically meaningful, that (i) amino acid sequence alignment ($r_{max} = 19$) is preferred, rather than nucleotide sequence alignment ($r_{max} = 3$), and that (ii) sequence sampling should be from a large phylogeny so that the real rank of mutational effects (r) can be reasonably close to r_{max}. For amino acid sequence alignment, the principle of microergodicity ensures that, when the phylogeny is sufficiently large, the rank (r) of mutational effects is close to the maximum value $r_{max} = 19$, leading to $K \approx \min(19, P_{min})$.

5.5 Case studies

5.5.1 Range of K_e estimated from protein sequences

One impressive prediction from our theory is that, based on the protein sequence alignment, the Gu (2007) method can estimate the degree of gene pleiotropy effectively only between 1 and 19, whereas this estimate would converge to 19 when the real gene pleiotropy becomes higher. We have examined three data sets (Su et al. 2010): 321 vertebrate genes (the phylogeny as shown by Fig. 5.3), 580 single-copy genes from five yeast genomes, and 437 single-copy genes from twelve Drosophila genomes. Although each data set shows a great range of gene pleiotropy, almost all estimates (99%) are < 19 (Table 5.1), consistent with the theoretical prediction.

Note: Effective gene pleiotropy (K_e) was estimated based on the human–mouse orthologous genes for d_N/d_S and Gu and Zhang's (1997) method for the H-measure. Biological processes were counted from the gene ontology (GO). Expression broadness is the number of mouse tissues in which a gene is expressed, based on the Su et al. (2004).

When the total evolutionary time of the phylogeny is not large enough to meet the criterion of microergodicity, the real rank (r) of mutational effects could be $<< r_{max}$, resulting in a lower rank (K) of the genotype-phenotype map. We tested this prediction by the vertebrate gene data set (Su et al. 2010). For each gene, we estimated Ke from four species (mammals), five species (mammals and chicken), six species (mammals, chicken, and frog), seven species (mammals, chicken, frog, and fugu), and eight species (mammals, chicken, frog, fugu, and zebrafish), respectively. We observed the average of $K_e = 3.1$, 4.2, 6.0, 6.3, and 6.5 for the data sets of four species to eight species, respectively. Indeed, for the same gene set, the estimation of K_e tends to be small when the protein phylogeny is small. One may wonder about the ultimate value that Ke may approach when the phylogeny becomes very large. Tentatively, we used a simple extrapolation model after calculating the total branch length of the phylogeny in each case and estimated and found that, roughly, the average of K_e approaches 7.17.

5.5.2. K_e estimated from nucleotide sequences reflecting the rank (r) of mutational effects

Estimation procedure for nucleotide sequences

We first calculated the quantity $g = (d_{nuc}/d_S)/(1 - H)$ for each gene. Here d_{nuc} is the evolutionary distance at the first and second nucleotide positions (d_{12}) or at all three positions (d_{123}), and d_S is the synonymous distance. All these distances were estimated between human and mouse, by the Kimura two-parameter method. When the phylogeny is given, the normalized index of rate variation among nucleotide sites (H) can be calculated by the algorithm of Gu and Zhang (1997) with some modifications. Due to the saturated synonymous substitutions at many nucleotide sites, estimation of the quantity g for each gene could be subject to a large sampling variance. Alternatively, we implemented the following procedure: First, calculate the mean of g over all genes, as well as the confidence interval (25% and 75% quantiles, respectively); and, second, use the mean of g to estimate the mean of K_e, as well as the 25% and 75% quantiles, respectively.

Table 5.1 *Summary of vertebrate gene pleiotropy analysis. (From Su et al. 2010.)*

K K_e	Num	dN/dS	dN	dS	H	S	B_0	Biological processes	Expression broadness
<3	26	0.269±0.022	0.161±0.018	0.609±0.033	0.482±0.033	-6.89±0.76	3.38±0.49	1.62±0.28	6.10±2.61
3-4	29	0.184±0.010	0.112±0.008	0.607±0.030	0.496±0.025	-7.14±0.24	2.08±0.06	1.76±0.35	5.73±2.76
4-5	47	0.134±0.006	0.090±0.006	0.651±0.029	0.492±0.022	-8.30±0.23	1.83±0.05	1.53±0.23	2.26±0.55
5-6	63	0.103±0.003	0.072±0.004	0.684±0.031	0.483±0.016	-9.03±0.12	1.65±0.02	1.70±0.27	8.64±2.05
6-7	42	0.065±0.002	0.045±0.004	0.678±0.042	0.557±0.016	-10.59±0.14	1.62±0.02	1.74±0.31	8.36±2.41
7-8	37	0.051±0.002	0.034±0.003	0.651±0.043	0.537±0.018	-11.42±0.15	1.53±0.02	2.08±0.38	6.82±2.79
8-9	27	0.036±0.002	0.022±0.002	0.609±0.046	0.570±0.023	-12.63±0.20	1.50±0.02	2.85±0.80	10.70±3.77
9-10	19	0.026±0.002	0.018±0.002	0.668±0.062	0.559±0.031	-13.75±0.25	1.44±0.03	2.00±0.48	10.57±4.03
>10	31	0.010±0.001	0.006±0.001	0.579±0.044	0.611±0.027	-18.03±0.61	1.36±0.02	2.26±0.62	13.36±3.24

Table 5.2 *The means and ranges of effective gene pleiotropy at the first and second nucleotide positions, all three nucleotide positions, and amino acid sites*

	1+2 position	1+2+3 position	Amino acid sites
Mean of d/d_s	0.115	0.269	0.113
Mean of H	0.724	0.689	0.221
Mean of g	0.416	0.866	0.145
25% of g	0.186	0.690	0.089
75% of g	0.590	0.963	0.245
Mean of K	2.914	0.477	6.451
75% of K	5.587	1.233	8.042
25% of K	1.753	0.125	4.677

Vertebrate genes

Vertebrate genes used in this analysis were from the data set of Su et al. (2010). For each gene, the MSA of nucleotides was transformed exactly from the MSA of amino acids. According to our model, we have $K = \min(r, n) < 3$ because $r_{max} = 3$ at nucleotide sites. We used the vertebrate gene set (Su et al. 2010) to test this prediction. Two types of nucleotide sets were analyzed: The first data set includes the first and second codon positions (data set 1st-2nd), and the second one includes all sites (data set all). We analyzed 321 vertebrate single-copy genes and found that the average of K_{12} based on the first and second codon positions, is 2.91, with the 25% sample quantile ($1.75 \sim 5.59$). This result can be reasonably explained by $r_{max} = 3$ such that $K_{12} \approx \min(3, n) \approx 3$ despite that the real gene pleiotropy n may be much higher. Moreover, when all nucleotide positions are used, the average of effective gene pleiotropy (K_{123}) is as low as 0.48, with the 25% sample quantile ($0.00 \sim 1.23$). Indeed, synonymous substitutions mainly at the third codon position have little functional effect (Table 5.2).

References

Barton, N.H. (1990). Pleiotropic models of quantitative variation. *Genetics* 124, 773–782.

Chen, W.H., Su, Z.X., and Gu, X. 2013 A note on gene pleiotropy estimation from phylogenetic analysis of protein sequences. *Journal of Systematics and Evolution* 51, 365–369.

Chevin, L.M., Martin, G., and Lenormand, T. (2010) Fisher's model and the genomics of adaptation: restricted pleiotropy, heterogenous mutation, and parallel evolution. *Evolution* 64, 3213–3231.

Cooper, T.F., Ostrowski, E.A. and Travisano, M. (2007) A negative relationship between mutation pleiotropy and fitness effect in yeast. *Evolution* 61, 1495–1499.

Dudley, A.M., Janse, D.M., Tanay, A., Shamir, R., and Church, G. M. (2005) A global view of pleiotropy and phenotypically derived gene function in yeast. *Molecular Systems Biology* 1, 2005.0001.

Fisher, R. A. (1930) *The Genetical Theory of Natural Selection.* Oxford University Press, Oxford.

Gu, X. (2007) Evolutionary framework for protein sequence evolution and gene pleiotropy. *Genetics* 175, 1813–1822.

Gu, X. (2014). Pleiotropy can be effectively estimated without counting phenotypes through the rank of a genotype-phenotype map. *Genetics* 197, 1357–1363.

Gu, X., Fu, Y.X., and Li, W.H. (1995). Maximum likelihood estimation of the heterogeneity of substitution rate among nucleotide sites. *Molecular Biology and Evolution* 12, 546–557.

Gu, X., and Zhang, J. (1997) A simple method for estimating the parameter of substitution rate variation among sites. *Molecular Biology and Evolution* 14, 1106–1113.

Hartl, D.L., and Taubes, C.H. (1996) Compensatory nearly neutral mutations, selection without adaptation. *Journal* of *Theoretical Biology* 182, 303–309.

Hartl, D.L., and Taubes, C.H. (1998) Towards a theory of evolutionary adaptation. *Genetica* 102–103, 525–533.

Kimura, M. (1983) *The Neutral Theory of Molecular Evolution.* Cambridge University Press, Cambridge, UK.

Lande, R. (1980) The genetic covariance between characters maintained by pleiotropic mutations. *Genetics* 94, 203–215.

MacLean, R.C., Bell, G. and Rainey, P. B. (2004) The evolution of a pleiotropic fitness tradeoff in *Pseudomonas fluorescens. Proceedings of the National Academy of Sciences of the United States of America* 101, 8072–8077.

Martin, G., and Lenormand, T. (2006) A general multivariate extension of Fisher's geometrical model and the distribution of mutation fitness effects across species. *Evolution* 60, 893–907.

Martin, G., Elena, S.F., and Lenormand, T. (2007) Distributions of epistasis in microbes fit predictions from a fitness landscape model. Nat. Genet. 39, 555–560.

Paaby, A.B., and Rockman, M.V. (2013) The many faces of pleiotropy. *Trends in Genetics* 29, 66–73.

Poon, A., and Otto, S.P. (2000) Compensating for our load of mutations: freezing the meltdown of small populations. *Evolution* 54, 1467–1479.

Razeto-Barry, P., Diaz, J., Cotoras, D. and Vasquez, R.A. (2011) Molecular evolution, mutation size and gene pleiotropy: a geometric reexamination. *Genetics* 187, 877–885.

Razeto-Barry, P., Diaz, J., and Vasquez, R.A. (2012) The nearly neutral and selection theories of molecular evolution under the Fisher geometrical framework: substitution rate, population size, and complexity. *Genetics* 191, 523–534.

Su, Z., Zeng, Y. and Gu, X. (2010) A preliminary analysis of gene pleiotropy estimated from protein sequences. *Journal of Experimental Zoology: Part B Molecular Development and Evolution* 314, 115–122.

Turelli, M. (1985) Effects of pleiotropy on predictions concerning mutation-selection balance for polygenic traits. *Genetics* 111, 165–195.

Wagner, G.P. (1989) Multivariate mutation-selection balance with constrained pleiotropic effects. *Genetics* 122, 223–234.

Wagner, G.P., and Zhang, J. (2011) The pleiotropic structure of the genotype-phenotype map: the evolvability of complex organisms. *Nature Reviews Genetics* 12, 204–213.

Waxman, D., and J.R. Peck, (1998) Pleiotropy and the preservation of perfection. *Science* 279, 1210–1213.

Welch, J. J., and Waxman, D. (2003) Modularity and the cost of complexity. *Evolution* 57, 1723–1734.

Wingreen, N.S., Miller, J., and Cox, E.C. (2003) Scaling of mutational effects in models for pleiotropy. *Genetics* 164, 1221–1228.

Zhang, X.S., and Hill, W.G. (2003) Multivariate stabilizing selection and pleiotropy in the maintenance of quantitative genetic variation. *Evolution* 57, 1761–1775.

6

Evolution of Genetic Robustness after Gene Duplication

The role of functional compensation by duplicate genes has been examined in diverse organisms by comparing the proportion (P_E) of essential genes in duplicates to P_E in singletons (Wagner 2000; Gu et al. 2003; Conant and Wagner 2004; Hanada et al. 2009). Though the pattern of duplicate compensation is universal, the evolutionary pattern is complex (Gu 2022). When an essential gene is duplicated, duplicate compensation is the only mechanism to keep two duplicate copies dispensable. On the other hand, when a dispensable gene is duplicated, ancient genetic buffering and duplicate compensation together keep both dispensable. Gu (2022) developed a statistical model to estimate the proportion of essential genes that were duplicated from essential genes, and that from dispensable genes, respectively, providing some new insights into the evolutionary pattern of genetic robustness after gene duplication. We discuss the main results in this chapter.

6.1 A simple model of genetic robustness after gene duplication

6.1.1 Genetic robustness between duplicate genes

The mathematical notations and biological interpretations are summarized in Table 6.1. A gene is called "essential" (denoted by d^-) if the single-gene deletion phenotype is severe or lethal, or "dispensable" (denoted by d^+) if its deletion phenotype is normal or nearly-normal (Ihmels et al. 2007; Hsiao and Vitkup 2008; Su et al. 2014; Kabir et al. 2017; Cacheiro et al. 2020; see also Rancati et al. (2018) for a comprehensive review). Consider two paralogous genes (A and B) duplicated from a common ancestor (O) t time units ago. There are four combined states, denoted by (d_A, d_B), representing double-dispensable (d^+, d^+), semi-dispensable (d^+, d^-) or (d^-, d^+), or double-essential (d^-, d^-), respectively.

Statistical Analysis of Molecular and Genomic Evolution. Xun Gu, Oxford University Press. © Xun Gu (2024).
DOI: 10.1093/oso/9780198816515.003.0006

Table 6.1 *A summary of mathematical notations and biological interpretations*

Notation	Interpretation
d^+	State of '*dispensable*' if the single-gene deletion phenotype is normal
d^-	State of '*essential*' if the single-gene deletion phenotype is severer or lethal
O^+	Duplication of a dispensable gene (ancestral dispensability)
O^+-*duplicates*	Duplicates from ancestrally dispensable genes
O^-	Duplication of an essential gene (ancestral essentiality)
O^--*duplicates*	Duplicates from ancestrally essential genes
$Q(d_A, d_B \backslash O^+)$	Probability of duplicates A and B being (d_A, d_B) alter t time units since duplication, conditional of ancestral dispensability (O^+); $d_A, d_B = d^+$ or d^-
$Q(d_A, d_B \backslash O^-)$	Probability of duplicates A and B being (d_A, d_B) after t time units since duplication, conditional of ancestral essentiality (O^-); $d_A, d_B = d^+$ or d^-
$Q(d_A, d_B)$	Probability of duplicates A and B being (d_A, d_B) alter t time units since duplication; $d_A, d_B = d^+$ or d^-
R_O	Probability of a gene pair duplicated from a dispensable gene. ie., $R_O = P(O^+)$
P_E	Proportion of essential genes in duplicates
$P_E(O^+)$	Proportion of essential genes in O^+-duplicates
$P_E(O^-)$	Proportion of essential genes in O^--duplicates

Material from: Gu, dN/dS-H, a New Test to Distinguish Different Selection Modes in Protein Evolution and Cancer Evolution, Journal of Molecular Evolution, published 2022, Springer Nature

We are interested in the derivation of $Q_t(d_A, d_B)$, the probability of any joint states (d_A, d_B) at time t since the duplication. To this end, one should distinguish between the duplication of an essential gene (ancestral essentiality, denoted by O^-) and the duplication of a dispensable gene (ancestral dispensability, denoted by O^+). Let $Q_t(d_A, d_B|O^-)$ be the probability of being (d_A, d_B) after t time units since gene duplication, conditional of the ancestral essentiality (O^-), and $Q_t(d_A, d_B|O^+)$ be the probability conditional of the ancestral dispensability (O^+). Since the ancestral state (dispensable or essential) for a duplicate pair is usually unknown, a mixture model is then implemented: let $R_0 = P(O^+)$ be the probability of a gene pair duplicated from a dispensable gene, and $1 - R_0 = P(O^-)$ be that from an essential gene (Liang and Li 2007; Liao and Zhang 2007; Su and Gu 2008). Together, one can write

$$Q_t(d_A, d_B) = (1 - R_0)Q_t(d_A, d_B|O^-) + R_0 Q_t(d_A, d_B|O^+) \quad (6.1)$$

where $(d_A, d_B) = (d^+, d^+)$, (d^+, d^-), (d^-, d^+), or (d^-, d^-), respectively. Note that the process of non-functionalization of one duplicate copy was not conceptualized in the model, which is the most common fate of duplicated genes. This treatment is rational under the assumption that the rate of non-functionalization was the same between dispensable and essential genes before duplication (Stark et al. 2017).

6.1.2 Duplication of essential gene: the notion of sub-functionalization

When an essential gene was duplicated, the process of sub-functionalization at the gene regulation level has been thought to be the major evolutionary mechanism for dupli-cate preservation (Force et al. 1999; Stoltzfus 1999; Prince and Pickett 2002; Innan and Kondrashov 2010; Stark et al. 2017). As a result, both duplicate copies can be pre-served without invoking positive selection. Suppose a duplicate pair has m independent functional components, each of which is either "active" (denoted by "1") or "inactive" (denoted by "0"). Let U_{11} be the probability of a component being active in both genes; U_{01} (or U_{10}) is that of being inactive in gene A but active in gene B (or active in A but inactive in B); and U_{00} be the probability of a component being inactive in both genes. Without loss of generality, it is assumed that $U_{01} = U_{10}$. According to the not-all-inactive constraint, i.e., each component is functionally active in at least in one duplicate copy, we claim $U_{00} = 0$, leading to $U_{11} = 1 - 2U$ and $U_{10} = U_{01} = U$, respectively. That is, with a probability of $2U$, a functional component is active in one duplicate but inac-tive in another one, and with a probability of $1 - 2U$, a component is active in both duplicates.

If these functional components of a gene are statistically independent and identical, $Q_t(d_A, d_B | O^-)$ can be derived in terms of the component parameter (U) and the number (m) of functional components, that is,

$$Q(d^+, d^+ | O^-) = (1 - 2U)^m$$
$$Q(d^+, d^- | O^-) = (1 - U)^m - (1 - 2U)^m$$
$$Q(d^-, d^+ | O^-) = (1 - U)^m - (1 - 2U)^m$$
$$Q(d^-, d^- | O^-) = 1 - 2(1 - U)^m + (1 - 2U)^m \tag{6.2}$$

The rationale of Eq. (6.2) is follows. Under the m-component model ($m > 1$), two duplicate copies remain both dispensable only when each component is active in both duplicates (with a probability of $1 - 2U$), which leads to the derivation of $Q_t(d^+, d^+ | O^-)$ directly. Next we consider the (marginal) probability of dispensability (d^+) con-ditional of the ancestral essentiality (O^-), denoted by $Q_t(d^+ | O^-)$. It appears that $Q_t(d^+ | O^-) = (1 - U)^m$ because the probability of a component being active in one duplicate is given by $(1 - U)$. Since $Q_t(d^+|O^-) = Q_t(d^+, d^+|O^-) + Q_t(d^+, d^-|O^-)$, it is straightforward to obtain the second and third equations of Eq. (6.2). The last equation of Eq. (6.2) is derived by the sum of probabilities to be 1. In addition, Eq. (6.2) implies a gradual process of state transition. The starting states were apparently double-dispensable (d^+, d^+). Over time, most of them could be transformed to semi-dispensable (d^+, d^-), with some of these eventually becoming double-essential (d^-, d^-).

6.1.3   Duplication of dispensable genes: the notion of rare neo-functionalization

When a dispensable gene is duplicated, gene dispensability can be maintained through ancient genetic buffering and/or duplicate compensation (Prince and Pickett 2002; Innan and Kondrashov 2010; Stark et al. 2017). As a result, sub-functionalization becomes an ineffective approach for the retention of duplicate genes, because the process of complementation between functional components (Force et al. 1999) is difficult to achieve. To explain, one may consider a simple case that two duplicates A and B have two sub-functions (F_1 and F_2). After a complete sub-functionalization, duplicate A has functional F_1 and nonfunctional F_2, whereas duplicate B has nonfunctional F_1 and functional F_2. Since both sub-functions are required at the organismal level, duplicates A and B obviously become essential in the case of no genetic buffering. However, if duplicates A and B are from the duplication of a dispensable gene, the status of dispensability would not be altered.

While neo-functionalization has been suggested for the duplicate preservation in the case of genetic buffering (Chen et al. 2010; Vankuren and Long 2018; Lee and Szymanski 2021), it is unlikely that both copies acquire new functions simultaneously. In this sense we assume that

$$Q_t(d^-, d^-|O^+) = 0 \tag{6.3}$$

This assumption holds well except for very ancient duplicates that may acquire new functions in the later stage. One may reasonably argue that the retention of dispensable genes through neo-functionalization may be mainly driven by a positive selection with a nontrivial selection strength.

6.1.4   Analysis of genetic robustness model between duplicates

Model formulation and estimation

Together with Eq. (6.2) and Eq. (6.3), the model of genetic robustness between duplicates formulated by Eq. (6.1) can be further specified as follows

$$Q_t(d^+, d^+) = (1 - R_0)(1 - 2U)^m + R_0 \left[1 - 2Q(d^-|O^+)\right]$$
$$Q_t(d^+, d^-) = (1 - R_0)\left[(1 - U)^m - (1 - 2U)^m\right] + R_0 Q(d^-|O^+)$$
$$Q_t(d^-, d^+) = (1 - R_0)\left[(1 - U)^m - (1 - 2U)^m\right] + R_0 Q(d^-|O^+)$$
$$Q_t(d^-, d^-) = (1 - R_0)\left[1 - 2(1 - U)^m + (1 - 2U)^m\right] \tag{6.4}$$

where $Q_t(d^-|O^+)$ is the probability of an O^+-duplicate being essential (d^-); under Eq. (6.3), one can show $Q_t(d^-|O^+) = Q_t(d^-, d^-|O^+) + Q_t(d^-, d^+|O^+) = Q_t(d^-, d^+|O^+)$. Note that there are four unknown parameters, R_0, U, m, and $Q(d^-|O^+)$ in two independent equations. Gu (2022) implemented a practically feasible approach to solve this problem, as shown below.

(*i*) Suppose we have a set (N) of duplicate pairs; all $2N$ genes have single-gene deletion phenotypes (dispensable or essential). Three types of duplicate pairs are considered, that is, *DD* for (d^+, d^+), *DE* for (d^+, d^-) or (d^-, d^+), and *EE* for (d^-, d^-), and their frequencies are denoted by f_{DD}, f_{DE}, and f_{EE}, respectively.

(*ii*) R_0, the (prior) probability of a gene being dispensable before gene duplication can be replaced by the proportion of single-copy dispensable genes in the current genome as a proxy, under the assumption that R_0 remained a rough constant during long-term evolution (Su and Gu 2008).

(*iii*) The parameter U can be estimated by replacing $Q_t(d^-, d^-)$ in the last equation of Eq. (6.4) by f_{EE}, that is,

$$1 - 2(1 - \hat{U})^m + (1 - 2\hat{U})^m = \frac{f_{EE}}{1 - R_0} \tag{6.5}$$

where m, the number of functional components, is treated as a known integer, i.e., $m = 2, 3, \ldots$.

(*iv*) The proportion of essential O^--duplicates, i.e., those duplicated from essential genes, is given by $Q(d^-|O^-) = Q(d^-, d^-|O^-) + Q(d^-, d^+|O^-)$. When U is estimated by Eq. (6.5) (for any fixed m), according to Eq. (6.2) one can estimate $Q_t(d^-|O^-)$ by

$$\hat{Q}_t(d^-|O^-) = 1 - \left(1 - \hat{U}\right)^m \tag{6.6}$$

(*v*) After replacing $Q_t(d^+, d^+)$ in the first equation of Eq. (6.4) by f_{DD}, one can show that the proportion of essential O^+-duplicates can be estimated by

$$\hat{Q}_t(d^-|O^+) = \frac{1}{2} - \frac{f_{DD} - (1 - R_0)\left(1 - 2\hat{U}\right)^m}{2R_0} \tag{6.7}$$

In short, from the observed frequencies f_{DD}, f_{DE}, and f_{EE} with two degrees of freedom, we attempt to estimate two parameters $Q(d^-|O^-)$ and $Q(d^-|O^+)$ by Eqs. (6-5) to Eq. (6.7). To this end, we use the proportion of single-copy dispensable genes in the current genome as a proxy of R_0, and m as a constant that may only affect our estimation marginally.

Statistical evaluation

The statistical properties of two estimates, $Q(d^-|O^-)$ and $Q(d^-|O^+)$, can be evaluated by two approaches. First, their large-sample variances can be obtained by the delta-method under a multinomial model of f_{DD}, f_{DE}, and f_{EE}. The analytical formulas can be approximately obtained though the algebra is tedious. Second, a bootstrapping approach is implemented to empirically determine the sampling variance, as well as the confidence internals of these estimates.

Effect of the number of functional components (m)

The number of sub-functions (m) involved in the process of sub-functionalization after gene duplication actually represents a subset of subfunctions that are essential for the fitness of the organism. By computer simulations, Gu (2022) examined how the number (m) of functional components may affect our analysis. Note that the model of sub-functionalization requires at least two functional components. Hughes and Liberles (2007) suggested that between $m = 2$ and $m = 12$ regulatory regions would be biologically realistic. By extensive simulation analysis, Stark et al. (2017) argued that it was unlikely that a gene would have in excess of $m = 20$ functional components. The main simulation results are follows: (*i*) the estimate of $Q(d^-|O^-)$ tends to decrease slightly when m is increased from 2 to 5 (about 20%), whereas that of $Q(d^-|O^+)$ tends to increase slightly; (*ii*) in both cases little difference was observed for $m = 5$ or more; and (*iii*) all estimates are virtually the same from $m = 7$ to $m = 8$. In short, it seems that the effect of variable m is negligible as long as it is reasonably large, say, $m = 5$ or more.

Prediction of joint conditional probabilities

In practice it is desirable to know two types of conditional probabilities, $Q_t(d_A, d_B|O^-)$ and $Q_t(d_A, d_B|O^+)$, from the observed frequencies f_{DD}, f_{DE}, and f_{EE}. According to Eq. (6.2), it is straightforward to calculate the conditional probabilities of (d_A, d_B) after duplication of an essential gene (O^-) as

$$\hat{Q}_t(d^+, d^+|O^-) = (1 - 2\hat{U})^m$$

$$\hat{Q}_t(d^+, d^-|O^-) = (1 - \hat{U})^m - (1 - 2\hat{U})^m$$

$$\hat{Q}_t(d^-, d^+|O^-) = (1 - \hat{U})^m - (1 - 2\hat{U})^m$$

$$\hat{Q}_t(d^-, d^-|O^-) = \frac{f_{EE}}{1 - R_0} \tag{6.8}$$

where $\hat{U}$ is the positive solution of Eq. (6.5). Next, one can predict $Q(d^+, d^+|O^+)$ by equating $Q_t(d^+, d^+)$ with f_{DD} in Eq. (6.1) in the case of $d_A = d^+$ and $d_B = d^+$, and replacing $Q(d^+, d^+|O^-)$ by its prediction given by the first equation of Eq. (6.8). They are, respectively, given by

$$\hat{Q}_t(d^+, d^+|O^+) = \frac{f_{DD} - (1 - R_0)\left(1 - 2\hat{U}\right)^m}{R_0}$$

$$2\hat{Q}_t(d^+, d^-|O^+) = 1 - \frac{f_{DD} - (1 - R_0)\left(1 - 2\hat{U}\right)^m}{R_0}$$

$$\hat{Q}_t(d^-, d^-|O^+) = 0 \tag{6.9}$$

As indicated before, for a set of duplicate pairs with observed f_{DD}, f_{DE}, and f_{EE}, there are only two degrees of freedom. Hence, the statistical procedure described above treated R_0 and m as known constants and then estimated U and $Q(d^-|O^+)$. In this sense, Eq. (6.8) and Eq. (6.9) are not statistically well-justified to be treated as "estimates"; instead, they should be regarded as predicted values.

6.1.5 Case study: duplicate pairs from the whole genome duplication (WGD) in yeast or mouse

Data availability

Due to the different gene-silence/knockout technologies that are technically feasible, the criteria to determine gene essentiality or dispensability are usually not comparable between species such as yeasts and mice. Because fitness phenotypes after single-gene deletions were identified under experimental conditions, the population size under natural conditions would not affect the outcome.

In total 325 yeast duplicate pairs were collected, which were from the WGD (whole-genome duplication) in yeast about 100 million years ago (Kim and Yi 2006; Guan et al. 2007; Musso et al. 2008). According to common practice in yeast single-gene deletion genomics, the mean fitness of single gene-deletion for any yeast gene is measured by the growth rate of the strain with a single gene deleted relative to the average growth rate of wild strains in five growth media. Qualitatively, it can be further grouped into lethal, strong effect, moderate effect, and very weak effect (Gu et al. 2003). From the evolutionary view, a yeast gene is then classified as d^+ if it belongs to the very weak-effect group, or d^- otherwise. Under this classification, the proportion of dispensable single-copy genes (0.605) from Gu et al. (2003) is used as a proxy of R_0. One may wonder how the analysis would be affected by the binned fitness data. Actually, the fitness histogram showed a U-like pattern where the moderate-effect group is the smallest. In other words, our classification of yeast essential or dispensable genes should be robust against the bin cutoff.

The second dataset includes 217 mouse duplicate pairs from the WGD that occurred (Makino and McLysaght 2010) about 600 million years ago (in the early stage of vertebrates) (Wang and Gu 2000). Each pair was assigned by the mouse knockout phenotypes as follows (Su and Gu 2008). First, mouse phenotype and genotype association file was downloaded from Mouse Genome Informatics. Here an essential gene was defined as a gene whose knockout phenotype is annotated as lethality (including embryonic, prenatal, and postnatal lethality) or infertility. We excluded all the phenotypic annotations due to multiple gene knockout experiments, and only used those of null mutation homozygotes by target deletion or gene-trap technologies.

Analysis

Our analysis is focused on three variables: (*i*) P_E is the observed proportion of essential duplicates; (*ii*) $P_E(O-)$ is the expected proportion of essential O^--duplicates, i.e., those duplicated from essential genes, as estimated by $\hat{Q}(d^-|O^-)$ in Eq. (6.6); and (*iii*) $P_E(O^+)$ is the expected proportion of essential O^+-duplicates, i.e., those duplicated from dispensable genes, as estimated by $\hat{Q}(d^-|O^+)$ in Eq. (6.7). Their relationship is simply given by

$$P_E = (1 - R_0)P_E(O^-) + R_0 P_E(O^+) \qquad (6.10)$$

The frequencies of duplicate pairs with DD (double-dispensable), DE (dispensable-essential) and EE (double-essential) are presented in Fig. 6.1(A) (yeast) and Fig. 6.2(A) (mouse), respectively. While there is no empirical information about the number of functional components (*m*) for mouse and yeast genes, the robustness of the following analysis against various *m*-values is important. Consistent with the simulation result, our analysis was generally not affected by *m*; overall it revealed little difference among those cases of $m = 3$ or more. Our analysis of yeast WGD duplicate pairs is shown in Fig. 6.1(B), and that of mouse in Fig. 6.2(B) ($m = 6$). Roughly speaking, yeast WGD pairs represent the case of a recent WGD event, whereas mouse WGD pairs represent an ancient one. In the case of yeast WGD pairs, the proportion of essential duplicates ($P_E = 10.3\%$) is significantly larger than zero ($p < 10^{-6}$), yet it is much lower than

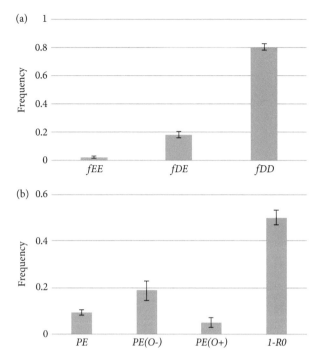

Figure 6.1 *Analysis of yeast 325 WGD pairs. (A) Frequencies of duplicate pairs with DD (double-dispensable), DE (dispensable-essential) and EE (double-essential) are presented. (B) The proportion of essential duplicates (P_E), the estimated P_E in O^--duplicates (duplication of essential genes), $P_E(O^-)$, and the estimated P_E in O^+-duplicates (duplication of dispensable genes), $P_E(O^+)$, are presented. In the analysis, the number of functional components is set to be m = 6. For comparison, the proportion of essential genes in single-copy genes (1-R_0) is also presented.*
Material from: Gu, dN/dS-H, a New Test to Distinguish Different Selection Modes in Protein Evolution and Cancer Evolution, Journal of Molecular Evolution, published 2022, Springer Nature

that of single-copy yeast genes ($P_{E,sin} = 39.5\%$). Gu (2022) showed that the P_E in O^--duplicates (duplication of essential genes) was $PE(O^-) = 21.2\%$, significantly greater than zero ($p < 0.001$), whereas P_E in O^+-duplicates (duplication of dispensable genes) is $P_E(O^+) = 3.0\%$ that was not significant ($p > 0.05$). Note that f_{EE} (the proportion of double-essential duplicate pairs) is so small that the estimation of U is subject to a large sampling variance. Nevertheless, it appears that the increase of P_E after the yeast WGD was mainly due to O^--duplicates, those duplicated from dispensable genes. Since the duplication time is the same for all duplicate pairs, one may predict that the rate of essentiality in O^--duplicates through sub-functionalization is about 7-fold (21.2/3.0) greater than that in O^+-duplicates through neo-functionalization.

In the case of mouse WGD pairs representing an ancient WGD, we observed $P_E = 62.2\%$, virtually the same as P_E in single-copy genes (Liang and Li 2007; Liao and Zhang 2007; Su and Gu 2008). As expected, the estimate of $P_E(O^-) = 86.0\%$

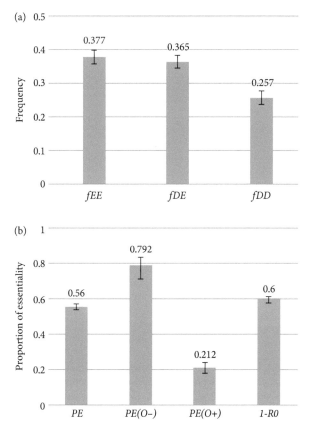

Figure 6.2 *Analysis of mouse 217 WGD pairs. (A) Frequencies of duplicate pairs with DD (double-dispensable), DE (dispensable-essential) and EE (double-essential) are presented. (B) The proportion of essential duplicates (P_E), the estimated P_E in O^--duplicates (duplication of essential genes), $P_E(O^-)$, and the estimated P_E in O^+-duplicates (duplication of dispensable genes), $P_E(O^+)$, are presented. In the analysis, the number of functional components is set to be $m = 6$. For comparison, the proportion of essential genes in single-copy genes (1-R_0) is also presented.*

Material from: Gu, dN/dS-H, a New Test to Distinguish Different Selection Modes in Protein Evolution and Cancer Evolution, Journal of Molecular Evolution, published 2022, Springer Nature

indicated that the majority of O^--duplicates in mouse WGD pairs, i.e., those dupli-cated from essential genes, may have become essential. Interestingly, the estimate of $P_E(O^+) = 28.4\%$ was significantly greater than zero ($p < 0.001$). Indeed, a nontrivial portion of O+-duplicates in mice, i.e., those duplicated from dispensable genes, may be essential, which were subjected to neo-functionalization after the gene duplication (Chen et al. 2010; Vankuren and Long 2018; Lee and Szymanski 2021).

   We observed that, strikingly, $P_E(O^-) > P_E(O^+)$ significantly in both WGD duplicate pairs ($p < 0.005$), which can be tentatively interpreted as follows: after the occurrence of WGD, the proportion of essential duplicates (P_E) increases with time t (the same for all

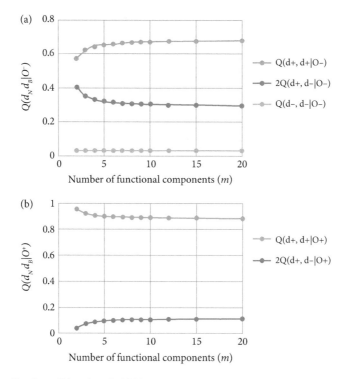

Figure 6.3 *Predicted conditional probabilities of yeast WGD duplicate pairs plotting against the number of functional components* $m = 2, \ldots, 20$. *(A)* $Q(d_A, d_B | O^-)$, *probabilities conditional of ancestral essentiality* (O^-). *(B)* $Q(d_A, d_B | O^+)$, *probabilities conditional of ancestral dispensability* (O^+).

Material from: Gu, dN/dS-H, a New Test to Distinguish Different Selection Modes in Protein Evolution and Cancer Evolution, Journal of Molecular Evolution, published 2022, Springer Nature

duplicate pairs) through two distinct evolutionary routes: a fast process of essentiality in O^--duplicates through sub-functionalization, and a slow process of essentiality in O^+-duplicates through neo-functionalization; the difference is about 3-fold (86/28.4). Finally, Fig. 6.3 shows the predicted conditional probabilities of yeast duplicate pairs: indeed, only marginal differences appeared when $m = 2$, and all estimates were virtually the same between $m = 5$ and $m = 8$. It was therefore concluded that the effect of variable m is usually negligible.

6.2 Evolutionary model of genetic robustness

6.2.1 Evolutionary model of O^--duplicates

When an essential gene was duplicated, the process of sub-functionalization, has been thought to be the major evolutionary mechanism for duplicate preservation (Force et al. 1999; Stoltzfus 1999; Prince and Pickett 2002; Innan and Kondrashov 2010; Stark et al. 2017). Consequently, both duplicate copies can be preserved without invoking positive selection. An evolutionary model for $Q_t(d_A, d_B|O^-)$, the probability of being (d_A, d_B) conditional of the ancestral essentiality (O^-), is proposed as follows.

Suppose a duplicate pair has m independent functional components. The evolution of every single component is described by a transition process between active ("1") to inactive ("0") states. Let U_{11} be the probability of a component being active in both genes; U_{01} (or U_{10}) is that of being inactive in gene A but active in gene B (or active in A but inactive in B); and U_{00} be the probability of a component being inactive in both genes. Let $P(i, j)$ be the probability that a single component has the status (i, j) in two duplicate genes A and B. The (two-state) Markov-chain model demonstrates that

$$P(1, 1) = \pi_1^2 + \pi_0 \pi_1 e^{-2\lambda t}$$

$$P(1, 0) = \pi_0 \pi_1 \left(1 - e^{-2\lambda t}\right)$$

$$P(0, 1) = \pi_0 \pi_1 \left(1 - e^{-2\lambda t}\right)$$

$$P(0, 0) = \pi_0^2 + \pi_0 \pi_1 e^{-2\lambda t} \tag{6.11}$$

where λ is the rate of state transition, π_1 (or π_0) is the stationary probability of a functional component being active (or inactive); $\pi_1 + \pi_0 = 1$. By the not-all-inactive constraint, i.e., each component is functionally active at least in one duplicate copy, the status $(0, 0)$ is not allowed. One may adjust these probabilities by $U_{00} = 0$ and $U_{ij} = P(i, j)/[1 - P(0, 0)]$ for others. From Eq. (6.11) it turns out to be $U_{10} = U_{01} = U$, and $U_{11} = 1 - 2U$, where U is given by

$$U = 1 - \frac{1}{1 + \pi_0(1 - e^{-2\lambda t})} \tag{6.12}$$

Apparently, $U_{11} = 1$ (and others are zero) at $t = 0$, and $U_{11} = \pi_1/(1 + \pi_0)$ and $U_{10} = U_{01} = \pi_0/(1 + \pi_0)$ as $t \to \infty$.

Note that the probability of an O^--duplicate to be essential is given by $Q(d^-|O^-) = 1 - (1 - U)^m$. Together with Eq. (6.12), we have

$$Q(d^-|O^-) = 1 - \left[\frac{1}{1 + \pi_0(1 - e^{-2\lambda t})}\right]^m \tag{6.13}$$

It appears that $Q(d^-|O^-) = 0$ at $t = 0$, then it increases toward a steady-state as $t \to \infty$. Let $\bar{Q}(d^-|O^-) = 1 - (1 + \pi_0)^{-m}$ be the steady-state probability of an O^--duplicate to be essential. Numerical analysis elaborated that Eq. (6.13) can be symbolically approximated by

$$Q(d^-|O^-) \approx \bar{Q}(d^-|O^-) \left(1 - e^{-2\mu t}\right) \tag{6.14}$$

where $\mu = m\lambda$ is the rate of an O^--duplicate to be essential after gene duplication.

6.2.2 Evolutionary model of O^+-duplicates

When a dispensable gene was duplicated, gene dispensability can be maintained through ancient genetic buffering and/or duplicate compensation (Prince and Pickett 2002; Innan and Kondrashov 2010). In this case, sub-functionalization is ineffective for duplicate gene preservation, because the process of complementation between functional components (Force et al. 1999) might be difficult to achieve. On the other hand, neo-functionalization was suggested for duplicate preservation that may break both genetic buffering and duplicate backup (Chen et al. 2010; Vankuren and Long 2018; Lee and Szymanski 2021). Nevertheless, it seems unlikely that both copies acquire new functions simultaneously. In this sense one may assume $Q(d^-, d^- | O^+) = 0$, which holds well except for very ancient duplicates that may acquire new functions in the later stage.

Let α be the rate of an O^+-duplicate to be essential after gene duplication. We presume that retention of dispensable genes through neo-functionalization is driven by positive selection. Hence, the magnitude of rate α is determined by the rate of adaptive mutation and the selection strength. Let $\bar{Q}(d^- | O^+)$ be the steady-state probability of an O^+-duplicate to be essential. Symbolically, the probability of an O^+-duplicate to be essential can be written as

$$Q(d^- | O^+) \approx \bar{Q}(d^- | O^+) \left(1 - e^{-2\alpha t}\right) \tag{6.15}$$

It appears that $Q(d^- | O^+) = 0$ at $t = 0$, and approaches to $\bar{Q}(d^- | O^+)$ as $t \to \infty$.

6.2.3 An approximate method of parameter estimation

Given frequencies of three types of duplicate pairs, *DD*, *DE*, and *EE*, we developed an approximate method that can facilitate the analysis. By Eq. (6.2) and Eq. (6.3), we have

$$
\begin{aligned}
Q(d^-, d^-) &= (1 - R_0) Q(d^-, d^- | O^-) \\
&= (1 - R_0)[1 - 2(1 - U)^m + (1 - 2U)^m] \\
&\approx (1 - R_0) \left[1 - (1 - U)^m\right]^2
\end{aligned} \tag{6.16}
$$

Numerical analysis shows that the last approximation holds well when m, the number of functional components, is reasonable large, such as $m > 5$ or more. Equating $Q_t(d^-, d^-)$ in Eq. (6.16) with the observed frequency f_{EE}, we have

$$\left[1 - (1 - \hat{U})^m\right]^2 \approx \frac{f_{EE}}{1 - R_0} \tag{6.17}$$

which provides a simple approach to estimating important parameters of genetic robustness, as shown below.

Let $P_E(O^-)$ be the expected proportion of essential O^--duplicates, i.e., those duplicated from essential genes, which can be estimated by $Q(d^-|O^+) = 1 - (1 - U)^m$. By Eq. (6.17), we obtain

$$P_E(O^-) = \hat{Q}(d^-|O^+) = \sqrt{\frac{f_{EE}}{1 - R_0}} \qquad (6.18)$$

where R_0, the (prior) probability of a gene being dispensable before gene duplication can be replaced by the proportion of single-copy dispensable genes in the current genome as a proxy, under the assumption that R_0 remained a rough constant during long-term evolution (Su and Gu 2008).

In the same manner, let $P_E(O^+)$ be the expected proportion of essential O^+-duplicates, i.e., those duplicated from dispensable genes, which can be estimated by

$$P_E(O^+) = \hat{Q}(d^-|O^+) = \frac{P_E - \sqrt{(1 - R_0)f_{EE}}}{R_0} \qquad (6.19)$$

6.2.4 Predicting double-gene knockout phenotypes of a duplicate pair

Double-deletion experimentation can predict the case of two dispensable duplicates maintained by their functional compensation if the cost to fitness of losing both duplicate genes is more severe than expected under the multiplicative model of genetic interaction (Costanzo et al. 2010; Guan et al. 2007). Hence, this negative genetic interaction would provide some convincing evidence for supporting the mechanism of duplicate compensation. Under the probabilistic model we formulated in this chapter, one may use Bayes rule to calculate the probability of ancestral essentiality (O^-) conditional of double-dispensability (d^+, d^+), that is,

$$Q_t(O^-|d^+, d^+) = \frac{(1 - R_0)Q_t(d^+, d^+|O^-)}{Q_t(d^+, d^+)} \qquad (6.20)$$

Therefore, the chance for a duplicate pair that are both dispensable to have a negative genetic interaction in the double-gene deletion experiments can be calculated as follows.

Let P_{negD} be the predicted proportion of dispensable duplicate pairs resulting in a negative genetic interaction when they are deleted simultaneously (or double-knockout), which can be estimated by calculating the conditional probability of ancestral essentiality (O^-) given by Eq. (6.20). After replacing $Q_t(d^+, d^+)$ in the first equation of Eq. (6.4) by f_{DD}, the (observed) proportion of double-dispensable duplicates, we have

$$f_{DD} = (1 - R_0)\hat{Q}(d^+, d^+|O^-) + R_0\hat{Q}(d^+, d^+|O^+)$$

$$= (1 - R_0)\hat{Q}(d^+, d^+|O^-) + R_0\left[1 - 2\hat{Q}(d^-|O^+)\right] \tag{6.21}$$

so that $\hat{Q}(d^+, d^+|O^-)$ can be written in terms of $\hat{Q}(d^-|O^+)$, f_{DD} and R_0. It follows that one can calculate $P_{negD} = \hat{Q}(O^-|d^+, d^+)$ as follows

$$P_{negD} = 1 - \frac{R_0[1 - 2\hat{Q}(d^-|O^+)]}{f_{DD}} \tag{6.22}$$

where $\hat{Q}(d^-|O^+)$ is given by Eq. (6.18).

6.3 Evaluation of Model Assumptions

The model for the evolution of genetic robustness is certainly oversimplified. First, essentiality and dispensability are relative categories for genes. In yeast, Hillenmeyer et al. (2008) found that 97% of gene deletions exhibited a measurable growth phenotype, suggesting that nearly all genes are essential for optimal growth under at least one condition. Therefore, the model of genetic robustness actually depends on a cutoff of fitness effect under a given environmental condition (Nowak et al. 1997; Visser et al. 2003; Flatt 2005). Indeed, dispensable genes in our case studies (yeast and mouse) should be interpreted as "nearly-dispensable" under ideal experimental conditions, whereas essential genes are likely to be truly "essential".

When an essential gene was duplicated, the current model assumed that two duplicate copies evolved under sub-functionalization, neglecting other possibilities such as neo-functionalization. Moreover, each functional component is assumed to undergo sub-functionalization independently, which is not biologically realistic (Szklarczyk et al. 2008; Hahn 2009; Chen et al. 2012; Keane et al. 2014; Saito et al. 2014; Diss et al. 2017; Teufel et al. 2018; Láruson et al. 2020; Mallik and Tawfik 2020). Meanwhile, after the duplication of a dispensable gene, interactions between ancestral genetic buffering, duplicate compensation and neo-functionalization remain largely unknown. In addition, some attributes of genetic mechanisms have not been taken into accounts, such as the effect of dosage balance, or the later-stage functional divergence (Prince and Pickett 2002; Innan and Kondrashov 2010).

A key assumption in our analyses is Eq. (6.3), that is, after duplication of a dispensable gene (O^+), the chance for both duplicate copies to be essential is negligible. While it is biologically intuitive, it may cause some bias, especially for some very ancient duplicate pairs. We conducted a simulation study to examine this effect. Our preliminary results showed that the estimation bias was usually marginal, except for an extremely long evolutionary span after gene duplication. In addition, the current model does not consider the neo-functionalization after the duplication of an essential gene if the acquired new function would not impair the current functions. Nevertheless, the neo-functionalization after sub-functionalization, or sub-neo-functionalization for short, would not change the status of essentiality.

References

Cacheiro, P., Muñoz-Fuentes, V., Murray, S.A., et al. (2020) Human and mouse essentiality screens as a resource for disease gene discovery. *Nature Communications* 11, 655.

Chen, S., Zhang, Y.E., and Long, M. (2010) New genes in Drosophila quickly become essential. *Science* 330, 1682–1685.

Chen, W.H, Trachana, K., Lercher, M.J., and Bork, P. (2012) Younger genes are less likely to be essential than older genes, and duplicates are less likely to be essential than singletons of the same age. *Molecular Biology and Evolution* 29, 1703–1706.

Conant, G.C., and Wagner, A. (2004) Duplicate genes and robustness to transient gene knockdowns in *Caenorhabditis elegans*. *Proceedings of the Royal Society Series B Biological Sciences* 271, 1534.

Diss, G., Gagnon-Arsenault, I., Dion-Coté, A.M., et al. (2017) Gene duplication can impart fragility, not robustness, in the yeast protein interaction network. *Science* 355, 630–634.

Flatt, T. (2005) The evolutionary genetics of canalization. *Quarterly Review of Biology*, 80, 287–316.

Force. A., Lynch. M., Pickett. F.B., et al. (1999) Preservation of duplicate genes by complementary, degenerate mutations. *Genetics* 151, 1531–1545.

Gu, X. (2003) Functional divergence in protein (family) sequence evolution. *Genetica* 118, 133–141.

Gu, X. (2022) A simple evolutionary model of genetic robustness after gene duplication. *Journal of Molecular Evolution* 90, 352–361.

Gu, Z., Steinmetz, L.M., Gu. X., et al. (2003) Role of duplicate genes in genetic robustness against null mutations. *Nature* 421, 63–66.

Guan. Y., Dunham, M.J., and Troyanskaya, O.G. (2007) Functional analysis of gene duplications in *Saccharomyces cerevisiae*. *Genetics* 175, 933–943.

Hahn, M.W. (2009) Distinguishing among evolutionary models for the maintenance of gene duplicates. *Journal of Heredity* 100, 605–617.

Hanada, K., Kuromori, T., Myouga, F., et al. (2009) Evolutionary persistence of functional compensation by duplicate genes in Arabidopsis. *Genome Biology and Evolution*, 1, 409–414.

Hillenmeyer, M.E., Fung, E., Wildenhain, J., et al. (2008) The chemical genomic portrait of yeast: Uncovering a phenotype for all genes. *Science* 320, 362–365.

Hsiao, T.-L., and Vitkup, D. (2008) Role of duplicate genes in robustness against deleterious human mutations. *PLoS Genetics* 4, e1000014.

Hughes, T., and Liberles, D.A. (2007) The pattern of evolution of smaller-scale gene duplicates in mammalian genomes is more consistent with neo- than subfunctionalisation. *Journal of Molecular Evolution* 65, 574–588.

Ihmels, J., Collins, S.R., Schuldiner, M., et al. (2007) Backup without redundancy: Genetic interactions reveal the cost of duplicate gene loss. *Molecular Systems Biology* 3, 86.

Innan, H., and Kondrashov, F. (2010) The evolution of gene duplications: Classifying and distinguishing between models. *Nature Reviews Genetics* 11, 97–108.

Kabir, M., Barradas, A., Tzotzos, G.T., et al. (2017) Properties of genes essential for mouse development. *PLoS ONE* 12, e0178273.

Keane, O.M., Toft, C., Carretero-Paulet, L., et al. (2014) Preservation of genetic and regulatory robustness in ancient gene duplicates of Saccharomyces cerevisiae. Genome Research 24, 1830–1841.

Kim, S.H., and Yi, S.V. (2006) Correlated asymmetry of sequence and functional divergence between duplicate proteins of *Saccharomyces cerevisiae*. *Molecular Biology and Evolution* 23, 1068–1075.

Láruson, Á.J., Yeaman, S., and Lotterhos, K.E. (2020) The importance of genetic redundancy in evolution. *Trends in Ecology and Evolution*, 35, 809–822.

Lee, Y., and Szymanski, D.B. (2021) Multimerization variants as potential drivers of neofunctionalization. *Science Advances* 7, : eabf0984.

Li, J., Yuan, Z., and Zhang, Z. (2010) The cellular robustness by genetic redundancy in budding yeast. *PLoS Genetics* 6, e1001187.

Liang, H., and Li, W.H. (2007) Gene essentiality, gene duplicability and protein connectivity in human and mouse. *Trends in Genetics* 23, 375–378.

Liang H, and Li, W.H. (2009) Functional compensation by duplicated genes in mouse. *Trends in Genetics* 25, 441–442.

Liao, B.Y., and Zhang, J. (2007) Mouse duplicate genes are as essential as singletons. *Trends in Genetics* 23, 378–381.

Makino, T., Hokamp, K., and McLysaght, A. (2009) The complex relationship of gene duplication and essentiality. *Trends in Genetics* 25, 152–155.

Makino, T., and McLysaght, A. (2010) Ohnologs in the human genome are dosage balanced and frequently associated with disease. *Proceedings of the National Academy of Sciences of the United States of America* 107, 9270–9274.

Mallik, S., and Tawfik, D.S. (2020) Determining the interaction status and evolutionary fate of duplicated homomeric proteins. *PLoS Computational Biology* 16, e1008145.

Musso, G., Costanzo. M., Huangfu, M.Q., et al. (2008) The extensive and condition-dependent nature of epistasis among whole-genome duplicates in yeast. *Genome Research* 18, 1092–1099.

Nowak, M.A., Boerlijst, M.C., Cooke, J., and Smith, J.M. (1997) Evolution of genetic redundancy. *Nature* 388, 167–171.

Prince, V.E., and Pickett, F.B. (2002) Splitting pairs: The diverging fates of duplicated genes. Nature Reviews Genetics 3, 827–837.

Rancati, G., Moffat, J., Typas, A., and Pavelka, N. (2018) Emerging and evolving concepts in gene essentiality. *Nature Reviews Genetics* 19, 34–39.

Saito, N., Ishihara, S., Kaneko, K. (2014) Evolution of genetic redundancy: The relevance of complexity in genotype-phenotype mapping. *New Journal of Physics* 16, 063013

Stark, T.L., Liberles, D.A., Holland, B.R., and O'Reilly, M.M. (2017) Analysis of a mechanistic Markov model for gene duplicates evolving under subfunctionalization. *BMC Evolutionary Biology* 17, 38.

Stoltzfus, A. (1999) On the possibility of constructive neutral evolution. *Journal of Molecular Evolution* 49, 169–181.

Su, Z., and Gu, X. (2008) Predicting the proportion of essential genes in mouse duplicates based on biased mouse knockout genes. *Journal of Molecular Evolution* 67, 705–709.

Su, Z., Wang, J., and Gu, X. (2014) Effect of duplicate genes on mouse genetic robustness: An update. *Advances in Computational Genomics* 2014, 758672.

Szklarczyk, R., Huynen, M.A., and Snel, B. (2008) Complex fate of paralogs. *BMC Ecology and Evolution* 8, 337.

Teufel, A.I., Johnson, M.M., Laurent, J.M., et al. (2018) Withdrawn as duplicate: The many nuanced evolutionary consequences of duplicated genes. *Molecular Biology and Evolution* 35, 304–314.

Vankuren, N.W., and Long, M. (2018) Gene duplicates resolving sexual conflict rapidly evolved essential gametogenesis functions. *Nature Ecology & Evolution* 2, 705–712.

Visser, J.A.G.M., Hermisson J, Wagner GP, et al. (2003) Perspective: evolution and detection of genetic robustness. *Evolution (NY)* 57, 1959–1972.

Wagner, A. (2000) Robustness against mutations in genetic networks of yeast. *Nature Genetics* 24, 355–361.

Wang, Y., and Gu, X. (2000). Evolutionary patterns of gene families generated in the early stage of vertebrates. *Journal of Molecular Evolution* 51, 88–96.

7

Statistical Models of Transcriptome Evolution

Many studies showed that variations in gene expression accounted for a substantial portion of phenotypic variations between species (Enard et al. 2002; Bergmann et al. 2003; Gu and Gu 2003; Rifkin et al. 2003; Khaitovich et al. 2005; Gilad et al. 2006b), supporting the notion that gene regulation may have played a key role in phenotypic evolution (King and Wilson 1975; Harrison et al. 2012; Lehner 2013). With the help of abundant transcriptome (RNA-seq) data from a broad range of organisms, large-scale comparative analysis has provided a power approach to investigating the underlying evolutionary mechanisms (Bedford and Hartl 2009; Brawand et al. 2011; Schraiber et al. 2013; Gallant et al. 2014; Pankey et al. 2014; Lamanna et al. 2015; Musser and Wagner 2015; Sudmant et al. 2015). To this end, an appropriate statistical framework for transcriptome evolution is highly desirable. While Brownian motion (BM) was suggested to describe neutral changes of gene expression (Khaitovich et al. 2004), further studies showed that the Ornstein-Uhlenbeck (OU) model that takes stabilizing selection into account should be used as a basic model for transcriptome evolution (Lemos et al. 2005; Gilad et al. 2006a; Gu and Su 2007; Bedford and Hartl 2009; Brawand et al. 2011; Rohlfs et al. 2014). Yet, due to the complexity of phenotypic evolution, it has been always controversial on how to interpret some fundamental evolutionary parameters (Lemos et al. 2005). In this chapter, we will address these issues, with special reference to the transcriptome evolution between species.

7.1 Stabilizing selection of transcriptome evolution and the OU model

For a given gene, suppose that the expression level can be quantitatively measured by a continuous, normal-like variable denoted by x. It has been assumed that the expression

Statistical Analysis of Molecular and Genomic Evolution. Xun Gu, Oxford University Press. © Xun Gu (2024).
DOI: 10.1093/oso/9780198816515.003.0007

level is subject to stabilizing selection around the optimum μ with the fitness function given by

$$f(x) = \exp\left\{-\frac{w(x-\mu)^2}{2}\right\}$$

(7.1)

where w is the coefficient of stabilizing selection; a large w means a strong selection pressure imposed on the expression level, and vice versa (Lemos et al. 2005; Bedford and Hartl 2009). According to the genetics model of phenotype evolution (Lande 1979), one can show that the first and second moments of the expression difference per generation are given by

$$E[\Delta x] = \sigma^2\frac{\partial \ln f}{\partial x} = \sigma^2 w(\mu - x)$$

$$E[\Delta^2 x] = \frac{\sigma^2}{2N_e}$$

(7.2)

respectively, where σ^2 is the rate of evolutionary variance. If the genotype-environment interaction and epistasis are negligible, $E[\Delta^2 x]$ mainly results from the sampling of heritable genotypic values of individuals, which is equal to the rate of evolutionary variance underlying the expression level (σ^2) divided by N_e, the effective population size, in the case of haploidy, or by $2N_e$ in the case of diploidy; to avoid confusion, diploidy is always assumed in the rest of analysis. Under the diffusion-limit approach, the OU stochastic process can be written as

$$dx(t) = a(x)dt + b(x)dB$$

(7.3)

(Hansen and Martins 1996), where the two parameters $a(x)$ and $b(x)$ are specified by

$$a(x) = \sigma^2 w(\mu - x)$$
$$b(x) = \sigma^2/(2N_e)$$

(7.4)

The first term on the right hand of Eq. (7.3) represents the deterministic selection pressure against the expression shift away from the optimum (μ), while the second term is the Brownian-type random mutational force (dB) to drive the expression level away from the optimum. When $\beta = 0$, the OU model is reduced to the BM model.

Let $\Psi(x, t)$ be the probability density of the expression level (x) at time t, which satisfies the following Kolmogrov forward equation

$$\frac{\partial \Psi(x, t)}{\partial t} = \frac{1}{2}\frac{\partial^2}{\partial x^2}\{b(x)\Psi(x, t)\} - \frac{\partial}{\partial x}\{a(x)\Psi(x, t)\}$$

(7.5)

Solving Eq. (7.5) results in the well-known OU distribution: given the initial frequency x_0, $\Psi(x, t)$ follows a normal distribution with the mean and variance given by

$$E[x|x_0] = x_0 e^{-\beta t} + \mu \left(1 - e^{-\beta t}\right)$$

$$V(x|x_0) = \frac{1}{W} \left(1 - e^{-2\beta t}\right) \tag{7.6}$$

where two important parameters to describe the pattern of transcriptome evolution are $\beta = \sigma^2 w$ and $W = 4N_e w$. As will be shown later, the rate of evolutionary variance per gene defined by

$$\eta^2 = \frac{\sigma^2}{2N_e} = \frac{2\beta}{W} \tag{7.7}$$

plays a critical role in the analysis of transcriptome evolution.

Further, one may define two limiting forms of the OU model: BM, which is the OU model with $\beta = 0$, and the white noise (WN) model, with $\beta = \infty$. Under BM, without selection, the expected expression level stays constant as time passes, given the initial value (x_0), but the variance $V(x|x_0) = \eta^2 t$ increases linearly with time t. The other extreme is the WN model: the rate of expression divergence is so fast that the expression level is expected to be almost immediately located around the optimum with a variance of $1/W$.

7.2 The OU model under a phylogeny

7.2.1 Phylogenetic comparative method (PCM) of transcriptome evolution

A simple case of PCM postulated that changes of expression level are deterministically attracted towards a single selective optimum (μ), at the rate β jointly determined by the strength of stabilizing selection (w) and the rate of evolutionary variance per gene (η^2). Recognizing that selection is a dynamic process, PCM has been extended to accommodate multiple selective optima. This model assumes various selective regimes across the phylogeny (e.g., Hansen 1997; Butler and King 2004). Under these selection themes, a strong stabilizing selection (a large w value) indicates a rapid rate of transcriptome evolution: the value of β determines the level of expression divergence along the phylogeny. BM, which is the OU model with $\beta = 0$, suggesting that the expected trait (expression) value of all current species given the root value stays constant as time passes. The other extreme is the WN model, which is the OU model with $\beta = \infty$, suggesting that all expression levels are expected to be almost immediately located around the optimum. In the OU model, the imprint of shared evolutionary history on species (the covariance between species) is generally progressively erased due to the effect of selection (Hansen and Martins 1996).

7.2.2 The stationary OU model of transcriptome evolution

The PCM interpretation of transcriptome evolution relies on a strong assumption of natural selection: the expression level at an ancestral node that was not at optimum evolved along the phylogeny toward the optimum. There is another evolutionary scenario we have to consider: what if the expression level at an ancestral node has been already at the optimum. In this case, strong stabilizing selection (a large w value) indicates a low rate of transcriptome evolution. In this case BM, the OU model with $\beta = 0$, predicts a maximum rate of expression divergence, whereas the WN model, which is the OU model with $\beta = \infty$, predicts that almost no expression divergence: expression levels are expected to be always located around the optimum. To address this issue, we (Ruan et al. 2016; Gu et al. 2019; Yang et al. 2020) introduced the stationary OU model.

Consider the evolutionary process from the origin of the a particular tissue to the divergence between species (Fig. 7.1): the first theme is the evolutionary lineage from the tissue origin at node Z) to the root (node O) of the species tree, with τ time units; and the second theme is the divergence along a species phylogeny. Although the timing of tissue origin was ancient, it seems likely that the gene expression was driven by a positive selection toward the optima maintained by the OU model. Let $\bar{x}_0$ and V_0 be the mean and variance at the root (O). According to Eq. (7.6), they are given by

$$\bar{x}_0 = \mu(1 - e^{-\beta_0 \tau}) + z e^{-\beta_0 \tau}$$

$$V_0 = \frac{1 - e^{-2\beta_0 \tau}}{W_0} \tag{7.8}$$

where z is the initial expression level at node Z, and β_0 and W_0 is the OU-rate and the strength of stabilizing selection in the ancient lineage, respectively. As τ is sufficiently large, one may anticipate that at the root of species phylogeny (node O), the OU process has virtually approached the stationary condition as if $\tau \to \infty$, so that the mean and the variance of x_0 at the root of phylogeny are simply given by μ and $1/W_0$. Since then, both the optimal level (μ) and the strength of stabilizing selection (W_0, referred to W thereafter) remained constant along the species phylogeny. Consequently, the expression variances in all internal and external nodes were the same ($1/W$). It has been shown that the stationary OU process along a phylogeny is root independent (Hansen and Martins 1996).

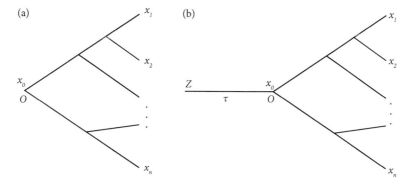

Figure 7.1 *(A) The evolutionary scenario for comparative transcriptome analysis. Given the expression level (x_0) at root O of the species phylogeny, the joint expression density of $x = (x_1, \ldots x_n)$ is given by $P(x|x_0)$, which is multi-variate normal distributed. Meanwhile, the evolutionary lineage from the origin of the tissue (nodeZ) to the root (node O), with τ evolutionary time units. Given the initial value z_0 at node Z, the OU process of x_0 along the lineage from Z to O is given by $OU(x_0|z_0; \mu, \beta\tau, W)$. When the tissue origin is so ancient that $\tau \to \infty$, it is called the stationary OU model along the species phylogeny. (B) The coefficient of expression correlation plotted against the evolutionary time (t) under the stationary OU model.*

7.3 Estimation of expression distance

7.3.1 Expression covariance between two species

Consider a simple scenario of transcriptome evolution with two species that diverged t time units ago. Let x_1 and x_2 be expression levels of two orthologous genes, respectively. Hansen and Martins (1996) showed that the expression covariance between x_1 and x_2 under the OU model is given by

$$Cov(x_1, x_2) = V_0 e^{-(\beta_1 + \beta_2)t} \tag{7.9}$$

where β_1 and β_2 are the rates of transcriptome evolution in lineage-1 and lineage-2, respectively, and V_0 is the variance at the common ancestor (node O).

Eq. (7.9) reveals that expression covariance between two species decays exponentially with the evolutionary time t, which can be used to estimate the expression distance defined by $D_{12} = 2\beta t$, where $\beta = (\beta_1 + \beta_2)/2$ is the mean evolutionary rate. However, the difficulty is that V_0, the ancestral variance at the root, is usually unknown. We (Ruan et al. 2017; Yang et al. 2020) invoked the stationary OU model to solve this problem: $V_0 = 1/W$. As the strength of stabilizing selection (W) remains constant along the species evolution, we have $Var(x_1) = Var(x_2) = V_0 = 1/W$. Let

$r_{12} = Cov(x_1, x_2)/\sqrt{Var(x_1)Var(x_2)}$ be the Pearson coefficient of correlation. Under the stationary OU model, Eq. (7.9) is simplified as follows

$$r_{12} = e^{-2\beta t} \tag{7.10}$$

7.3.2   The constant-μ method

Suppose we have N orthologous gene pairs, whose expression levels are denoted by x_{1k} and x_{2k}, $k = 1, \ldots, N$. The Pearson coefficient of (expression) correlation between two species is then calculated, denoted by $\hat{r}_{12}$. Under the assumption that the expression optimum (μ) is the same among genes, it is straightforward that this constant-μ expression distance can be simply estimated by

$$D_{12} = -\ln(1 - P_{12}) \tag{7.11}$$

where $P_{12} = 1 - \hat{r}_{12}$ is called the Pearson distance in the literature.

7.3.3   The variable-μ method

Since the expression optimum (μ) apparently varies among genes, the μ-constant assumption is biologically unrealistic. Indeed, computer simulations showed that neglecting the μ-variation among genes may lead to an underestimation of D_{12} by Eq. (7.11); the estimation bias becomes nontrivial when the expression distance is large, say, $D_{12} > 0.5$. Yang et al. (2020) developed a new method called the variable-μ method to correct this bias.

Suppose that the optimal expression value μ varies among genes according to a normal distribution with the mean zero and the variance V_μ. Under the stationary OU model, one can show that the expression variances of two orthologous genes are expected to be $Var(x_1) = Var(x_2) = 1/W + V_\mu = V_T$, and their covariance by

$$Cov(x_1, x_2) = \frac{e^{-2\beta t}}{W} + V_\mu \tag{7.12}$$

From the definition of Pearson coefficient of correlation r_{12}, we have

$$r_{12} = \frac{e^{-2\beta t}/W + V_\mu}{1/W + V_\mu} = \pi + (1 - \pi)e^{-2\beta t} \tag{7.13}$$

where $\pi = V_\mu/V_T$. It follows that the expression distance $D_{12} = 2\beta t$ can be estimated by

$$D_{12} = -\ln\left(\frac{\hat{r}_{12} - \pi}{1 - \pi}\right) = -\ln\left(1 - \frac{P_{12}}{1 - \pi}\right) \tag{7.14}$$

where π measures the degree of μ-variation among genes: $\pi = 0$ means a constant-μ assumption, whereas $\pi \to 1$ means a very strong μ-variation among genes.

7.3.4 Some examples

We estimated $\pi = 0.35 \sim 0.40$ based on mammalian RNA-seq data of six tissues (Brawand et al. 2011). In particular, Table 1 in Brawand et al. (2011) presented a detailed analysis of transcriptome evolution between human and macaque. The coefficient of expression correlation between species ranges from 0.71 to 0.90 among six tissues. We then estimated the expression distance D_{12} by three methods, i.e., the Pearson distance, the constant-μ distance, and the variable-μ distance. For instance, D_{12} of liver tissue is 0.104, 0.110 and 0.191, respectively, illustrating that different estimation methods may result in as many as two-fold differences. If one assumes that the human-macaque split time is about 29 million years (myr) ago, the rate of expression evolution in liver is around $1.79 \times 10^{-9} \sim 3.29 \times 10^{-9}$ per year.

7.4 Genetic basis of transcriptome evolution: fixed optimum model

In this section we briefly discuss an important issue about the genetic basis of OU-based transcriptome evolution between species under the fixed optimum model. Stabilizing selection has a dual-role on the rate of transcriptome evolution, depending on whether the initial state of gene expression is optimal or not.

7.4.1 Initial expression state is optimal: nearly-neutral evolution

In this case, any mutation is going to push away the gene expression from the optimum; the process is constrained by the stabilizing selection. Therefore, transcriptome evolution should be considered as the fixation process of neutral or nearly-neutral mutations into a population, subjecting to a weak purifying selection determined by the coefficient of stabilizing selection (w).

The quasi-fixation approximation

Note that for transcriptome evolution across species, the evolutionary variance (σ^2) represents the effect of mutations are fixed in the population. We propose a quasi-fixation approximation to address this issue, postulating that any expression shift $\delta = x - \mu$ of a gene is effectively caused by a single mutation, which is subject to a purifying selection since the expression optimum (μ) is fixed. Under the stabilizing selection model shown in Eq. (7.1), the coefficient of selection imposed on this effective mutation is given by $s = f(x) - 1 \approx -w\delta^2/2$ which is always negative. Let $u(\delta^2)$ be the fixation probability of an effective mutation that leads to the expression shift of δ, given the initial frequency $1/(2N)$, where N is the consensus population size. Under the Wright-Fisher model with additive dominance, $u(\delta^2)$ can be analytically solved by the diffusion model (Kimura 1983), that is,

$$u(\delta^2) \approx \left(\frac{1}{2N}\right)\left(\frac{2N_e s}{1 - e^{-2N_e s}}\right) = \left(\frac{1}{2N}\right)\left(\frac{2N_e w\delta^2}{e^{2N_e w\delta^2} - 1}\right) \tag{7.15}$$

where N_e is the effective population size. Apparently, the fixation probability of a mutation inversely depends on the quantity $2N_e w\delta^2$.

Rate of evolutionary variance per gene (η^2)

Technically one may treat the expression shift δ as a random variable that follows a normal distribution $\phi(\delta)$ with the mean 0 and the variance κ^2. For the interspecies transcriptome evolution is then defined by the quasi-fixation probability averaged by the expression shifts (δ) multiply the expected amount of new effective mutations per generation ($2Nv$). We therefore have

$$\eta^2 = 2Nv \int_{-\infty}^{\infty} \delta^2 u(\delta^2)\phi(\delta)d\delta$$
$$= v \int_{-\infty}^{\infty} \delta^2 \left(\frac{2N_e w\delta^2}{e^{2N_e w\delta^2} - 1}\right)\phi(\delta)d\delta \tag{7.16}$$

We further assume a normal distribution of $\phi(\delta)$ with mean 0 and variance κ^2. Using an approximation of $y/(e^y - 1) \approx e^{-y}$, we showed that $\sigma^2 = 2N_e\eta^2$ can be approximated by

$$\eta^2 \approx \frac{v\kappa^2}{(1 + W\kappa^2)^{3/2}} = \frac{\eta_0^2}{(1 + W\kappa^2)^{3/2}} \tag{7.17}$$

where $\eta_0^2 = v\kappa^2$ be the rate of mutational variance and $W = 4N_e w$ be the strength of stabilizing selection. Note that Eq. (7.17) implies an inverse relationship between η^2 and W: the rate of evolutionary variance is low for those genes under strong stabilizing selections in a large population, as expected by the nearly-neutral theory. It appears that $\eta^2 = \eta_0^2$ when $w = 0$ (no stabilizing selection), and $\eta^2 < \eta_0^2$ when $w > 0$.

The OU rate of transcriptome evolution (β)

According to the relationship given by Eq. (7.7), it is straightforward to have the OU rate of transcriptome evolution ($\beta = \sigma^2 w = 2N_e\eta^2 w$) as follows

$$\beta \approx \left[\frac{2N_e v\kappa^2}{(1 + W\kappa^2)^{3/2}} \right] w = \left[\frac{W\kappa^2}{2(1 + W\kappa^2)^{3/2}} \right] v \tag{7.18}$$

Interestingly, the β/v-W plotting can divide two themes: the near-BM theme when $W\kappa < 1$ and the OU theme when $W\kappa > 1$.

7.4.2 Initial expression state is not optimal: positive selection

In this case, gene expression evolves toward the optimum from the current expression value, driven by a positive selection whose selection coefficient is determined by the expression difference δ and the coefficient of stabilizing selection (w). According to Eq. (7.1), we approximately have $s \approx w\delta^2/2 > 0$. Let $U(\delta^2)$ be the fixation probability of the mutation, given the initial population frequency $1/(2N)$. By the quasi-fixation approximation, we have

$$U(\delta^2) \approx \left(\frac{1}{2N} \right) \left(\frac{2N_e w\delta^2}{1 - e^{-2N_e w\delta^2}} \right) \tag{7.19}$$

Similar above, we treat the expression shift δ as a random variable that follows a normal distribution $\Phi(\delta)$ with the mean 0 and the variance κ^2. It follows that the rate of evolutionary variance is given by

$$\eta^2 = 2Nv \int_{-\infty}^{\infty} \delta^2 U(\delta^2) \Phi(\delta) d\delta$$
$$= v \int_{-\infty}^{\infty} \delta^2 \left(\frac{2N_e w\delta^2}{1 - e^{-2N_e w\delta^2}} \right) \Phi(\delta) d\delta \tag{7.20}$$

After some approximations, we show that the rate of evolutionary variance per gene as follows

$$\eta^2 \approx v\kappa^2 \left[\frac{W\kappa^2/2}{1 - (1 + W\kappa^2)^{-1/2}} \right]$$
$$= \eta_0^2 \left[\frac{W\kappa^2/2}{1 - (1 + W\kappa^2)^{-1/2}} \right] \tag{7.21}$$

It is straightforward to derive the OU-rate of transcriptome evolution, that is

$$\beta \approx 2N_e v\kappa^2 \left[\frac{W\kappa^2/2}{1-(1+W\kappa^2)^{-1/2}} \right] w$$

$$= \left[\frac{(W\kappa^2/2)^2}{1-(1+W\kappa^2)^{-1/2}} \right] v \qquad (7.22)$$

7.5 Genetic basis of transcriptome evolution: random-shifted optimum

7.5.1 Random model of shifted optimum

Consider the OU model when the optimum (μ) changes randomly during the course of transcriptome evolution. Without loss of generality, suppose that, at the initial stage, the population mean of expression level is optimal in fitness. After the optimum is shifted to μ, the fitness function for any expression level (x) is given by $w(x) = e^{-w(x-\mu)^2/2}$. Next we assume that the fitness optimum μ varies according to a normal distribution $\psi(\mu)$ with the mean 0 and the variance σ_μ^2.

7.5.2 The coefficient of selection under the random-shifted optimum

First we show that given the optimum (μ), the coefficient of selection for a mutation with the expression level (x) can be written as

$$s(x|\mu) = \frac{w(x)}{w(\mu)} - 1 = \exp\left\{ -\frac{w}{2}\left[(x-\mu)^2 - \mu^2 \right] \right\} - 1 \qquad (7.23)$$

By adopting Taylor expanding $e^{-y} \approx 1 - y + y^2/2$ and then omitting those terms with higher orders than x^2 or μ^2, we obtain

$$s(x|\mu) \approx -\frac{w}{2}\left[(x-\mu)^2 - \mu^2 \right] + \frac{1}{2}\left\{ \frac{w}{2}\left[(x-\mu)^2 - \mu^2 \right] \right\}^2$$

$$\approx (\mu w)x - \frac{1}{2}w\left(1 - \mu^2 w \right) x^2 \qquad (7.24)$$

Under the random model optimum-shifts, one can show that

$$s(x) = \int_{-\infty}^{\infty} s(x|\mu)\psi(\mu)d\mu$$

$$= x^2 \left(\frac{w}{2} \right) \left(\sigma_\mu^2 w - 1 \right) = x^2 \left(\frac{w}{2} \right) (\gamma - 1) \qquad (7.25)$$

where $\gamma = \sigma_\mu^2 w$. It appears that $s(x) < 0$ when $\gamma < 1$, indicating a dominant role of stabilizing selection. By contrast, $s(x) > 0$ when $\gamma > 1$, indicating a dominant role of positive selection that results in the optimum shift.

7.5.3   Rate of evolutionary variances

Let $u(x)$ be the fixation probability of an effective mutation leading to an expression shift of x, given the initial frequency $1/(2N)$, where N is the consensus population size. Under the Wright-Fisher model with additive dominance, one can show that

$$u(x) \approx \left(\frac{1}{2N}\right)\left[\frac{4N_e s(x)}{1 - e^{-4N_e s(x)}}\right]$$
(7.26)

Similar to the case of fixed optimum, the expression shift x is treated as a random variable that follows a distribution denoted by $\phi(x)$ with the mean 0 and variance κ^2. Given the expected amount of new mutations (per generation) by $2Nv$, we obtain the rate of evolutionary variance per gene

$$\eta^2 = 2Nv \int_{-\infty}^{\infty} x^2 u(x)\phi(x)dx$$
$$= v \int_{-\infty}^{\infty} x^2 \left[\frac{4N_e s(x)}{1 - e^{-4N_e s(x)}}\right]\phi(x)dx$$
(7.27)

With the approximations similar to above, we show that

(*i*) If $\gamma < 1$, we have

$$\eta^2 = \frac{v\kappa^2}{[1 + W\kappa^2(1 - \gamma)]^{3/2}}$$
(7.28)

(*ii*) if $\gamma = 1$, we have

$$\eta^2 = v\kappa^2$$
(7.29)

(*iii*) if $\gamma > 1$, we have

$$\eta^2 = v\kappa^2 \frac{W\kappa^2(\gamma - 1)}{1 - [1 + W\kappa^2(\gamma - 1)]^{-1/2}}$$
(7.30)

7.6   Advanced topics: a unified framework of transcriptome evolution

7.6.1   The η^2-distance of transcriptome evolution

While the transcriptome distance defined by $D = 2\beta t$, referred to as β-distance there-after, is technically convenient to estimate, the analysis for its genetic basis reveals some difficulties about how to relate the observed rate of transcriptome evolution to the under-lying mechanisms. This is mainly because the evolutionary rate β is valid only under the OU model, which vanishes in the case of BM, i.e., the OU model with $w = 0$. Instead, our analysis showed that the rate of evolutionary variance per genes (η^2) may be more appro-priate to define the actually realized rate of transcriptome evolution. The η^2-distance of transcriptome evolution is defined as

$$E = 2\eta^2 t = 2 \left(\frac{2\beta}{W} \right) t \tag{7.31}$$

Let $E_0 = 2v\kappa^2 t$. Similar to the fundamental rule of molecular evolution, we claim

(*i*)   The η^2-distance satisfies $E > E_0$ when $\gamma > 1$, indicating adaptive evolution under the stabilizing selection

(*ii*)   The η^2-distance satisfies $E = E_0$ when $\gamma = 1$ or $w = 0$, indicating a quasi-neutrality or strict neutrality

(*iii*)   The η^2-distance satisfies $E < E_0$ when $\gamma < 1$, indicating a nearly-neutral evolu-tion under the stabilizing selection

7.6.2   Phylogenetic estimate of W

Suppose we have RNA-seq datasets of a particular tissue from n species, and the expres-sion profile of each k-th gene are denoted by $\mathbf{x}_k = (x_{1k}, \ldots, x_{nk})$, $k = 1, \ldots, N$. As long as the species pair of interest is specified, the expression distance ($2\beta t$) for a reference gene set can be estimated by the methods described above. Moreover, we developed a simple method to estimate W as follows.

(*i*)   Calculate gene-k specific mean (μ_k) by a simple average over orthologous genes.

(*ii*)   Calculate the matrix of correlation coefficients (**R**), i.e., R_{ij} is the coefficient of expression correlation between species i and j, and $R_{ii} = 1$.

(*iii*)   Let $\mathbf{C} = \mathbf{R}^{-1}$ and c_{ij} be the ij-th element of $\mathbf{C}$. Calculate the quadratic function of each gene k, $Q(\mathbf{x}_k)$, by

$$\hat{Q}(\mathbf{x}_k) = \sum_{i=1}^{n} \sum_{j=1}^{n} c_{ij}(x_{ik} - \mu_k)(x_{jk} - \mu_k) \tag{7.32}$$

where x_{ik} or x_{jk} is the expression value of gene k in species i or j, respectively.

(*iv*) The generalized linear-squared estimate of W is then given by

$$\hat{W} = \frac{n}{\hat{Q}(\mathbf{x}_k)} \tag{7.33}$$

7.6.3 Estimation of E_0: a tentative procedure

The most challenging problem to test the selection mode of transcriptome evolution is no golden standard for the neutral transcriptome evolution as a null hypothesis, that is, E_0 is usually difficult to estimate from the current transcriptome data. Tentatively, we propose two approaches to addressing this problem.

The reference gene-set approach

This approach requires two or more reference gene sets available with the assumption that E_0 and κ^2 are the same for all reference gene sets. Moreover, those gene sets as reference, such used as "house-keeping" genes, can be reasonably assumed to have little optimum shift during the transcriptome evolution ($\gamma < 1$ and close to 0).

Consider two given species under study called the target species pair. Suppose we have several (m) gene-set references. For each gene-set, we calculate the β-distance $D_k = 2\beta_k t$ and the strength of stabilizing selection (W_k), based on which the η^2-distance (E_k) is estimated, $k = 1, \ldots, m$. Draw the E_k-W_k plotting and then numerically fit the formula

$$E = \frac{E_0}{[1 + W\kappa^2(1 - \gamma)]^{3/2}} \tag{7.34}$$

using a (nonlinear) least-squared method to estimate E_0 and κ^2. Intuitively the estimate of E_0 can be obtained by extrapolating the curve to the point of $W = 0$.

It remains a challenge, in practice, to select reference gene sets objectively that are subject to the nearly neutral evolution during the transcriptome evolution. Tentatively, we may have two strategies. The first one is the biological approach. That is, selection of genes with known biological role that shown to be conserved during the transcriptome evolution. The second approach is the network approach to identifying genes within the co-expression modules during the species evolution, which will be addressed in future study.

The near-BM approach

Suppose that the strength of stabilizing selection (W) for each gene can be estimated by the empirical Bayesian approach (Gu et al. 2019), as will be discussed in detail in Chapter 9. We select two sets of genes based on the estimated β-distance for the given target species pair. The first one is for those gene with large β-distances, say, up 10% quantile. Let $\bar{W}_{up}$ be the sample mean of their W-estimates. It follows that, approximately, κ^2 can be estimated by

$$\hat{\kappa}^2 = \frac{1}{\bar{W}_{up}} \tag{7.35}$$

The second gene set is for those genes with low β-distances that is defined by $W < \bar{W}_{up}$. Let L be the number of those genes, considered as the nearly-BM theme. Hence, E_0 can be approximately estimated by

$$\hat{E}_0 = \frac{1}{L} \sum_{i=1}^{L} \frac{2D_i}{W_i} \tag{7.36}$$

7.6.4   Estimation of γ and quasi-neutrality test

We note that, under the optimum random-shift model of transcriptome evolution, the magnitude of selection intensity is given by

$$S = W(1 - \gamma) = 4N_e w(1 - \gamma) \tag{7.37}$$

The principle of quasi-neutrality test is to test $S = 0$, which means either $w = 0$ (strict neutrality, reduced to the BM model) or $\gamma = 1$ (quasi-neutrality under the OU model). To this end, a statistical test can be formulated to assert whether the rate of transcriptome evolution is significantly higher (or lower) than the neutral expectation. Specifically, the null hypothesis is $E = E_0$; alternatively, $E > E_0$ means $\gamma > 1$ (a strong adaptive evolution), or $E < E_0$ means $\gamma < 1$ (a nearly-neutral evolution with a weak or no micro-adaptation).

It is desirable to estimate the magnitude of γ and evaluate to what extent the expression optimum may have shifted randomly. From Eq. (7.34), it is straightforward to have

$$\gamma = 1 - \frac{1}{W\kappa^2} \left(\frac{E_0}{E} \right)^{\frac{2}{3}} \tag{7.38}$$

In practice, one may apply the genome-wide estimates of E_0 and κ^2 to achieve the goal after gene (gene-set)-specific estimates of W and E are available.

References

Bedford, T., and D.L. Hartl (2009) Optimization of gene expression by natural selection. *Proceedings of the National Academy of Sciences of the United States of America* 106, 1133–1138.

Bergmann, S., Ihmels, J. and Barkai, N. (2003) Similarities and differences in genome-wide expression data of six organisms. *PLOS Biology* 2, e9.

Blekhman, R., Oshlack, A., Chabot, A.E., Smyth G.K., and Gilad, Y. (2008) Gene regulation in primates evolves under tissue-specific selection pressures. *PLoS Genetics* 4, e1000271.

Brawand, D., Soumillon, M., Necsulea, A., et al. (2011) The evolution of gene expression levels in mammalian organs. *Nature* 478, 343–348.

Butler, M. A., and King, A.A. (2004) Phylogenetic comparative analysis, a modeling approach for adaptive evolution. *The American Naturalist* 164, 683–695.

Enard, W., Khaitovich, P., Klose, J., et al. (2002) Intra- and interspecific variation in primate gene expression patterns. *Science* 296, 340–343.

Gallant, J.R., Traeger, L.L., Volkening, J.D., et al. (2014) Nonhuman genetics. Genomic basis for the convergent evolution of electric organs. *Science* 344, 1522–1525.

Gilad, Y., Oshlack, A., and Rifkin, S.A. (2006a) Natural selection on gene expression. *Trends in Genetics* 22, 456–461.

Gilad, Y., Oshlack, A., Smyth, G.K., Speed, T.P., and White, K.P. (2006b) Expression profiling in primates reveals a rapid evolution of human transcription factors. *Nature* 440, 242–245.

Gu, J., and Gu, X. (2003) Induced gene expression in human brain after the split from chimpanzee. *Trends in Genetics* 19, 63–65.

Gu, X. and Su, Z. (2007) Tissue-driven hypothesis of genomic evolution and sequence-expression correlations. *Proceedings of the National Academy of Sciences of the United States of America* 104, 2779–2784.

Hansen, T.F., and Martins, E.P. (1996) Translating between microevolutionary process and macroevolutionary patterns: The correlation structure of interspecific data. *Evolution* 50, 1404–1417.

Harrison, P.W., Wright A.E., and Mank, J.E. (2012) The evolution of gene expression and the transcriptome–phenotype relationship. *Seminars in Cell & Developmental Biology* 23, 222–229.

Khaitovich, P., Hellmann, I., Enard, W.K., et al. (2005) Parallel patterns of evolution in the genomes and transcriptomes of humans and chimpanzees. *Science* 309, 1850–1854.

Khaitovich, P., Weiss, G., Lachmann, M., et al. (2004) A neutral model of transcriptome evolution. *PLoS Biology* 2, E132.

King, M., and Wilson, A. (1975) Evolution at two levels in humans and chimpanzees. *Science* 188, 107–116.

Lamanna, F., Kirschbaum, F., Waurick, I., Dieterich, C., and Tiedemann, R. (2015) Cross-tissue and cross-species analysis of gene expression in skeletal muscle and electric organ of African weakly-electric fish (Teleostei; Mormyridae). *BMC Genomics* 16, 668.

Lande, R., 1979 Quantitative genetic analysis of multivariate evolution, applied to brain: Body size allometry. *Evolution* 33, 402–416.

Lehner, B. (2013) Genotype to phenotype: Lessons from model organisms for human genetics. *Nature Reviews Genetics* 14, 168–178.

Lemos, B., Meiklejohn, C.D., Caceres, M., and Hartl, D.L. (2005) Rates of divergence in gene expression profiles of primates, mice, and flies: Stabilizing selection and variability among functional categories. *Evolution* 59, 126–137.

Musser, J.M., and Wagner, G.P. (2015) Character trees from transcriptome data: Origin and individuation of morphological characters and the so-called "species signal". *Journal of Experimental Zoology. Part B, Molecular and Developmental Evolution* 324, 588–604.

Pankey, M.S., Minin, V.N., Imholte, G.C., Suchard, M.A., and Oakley, T.H. (2014) Predictable transcriptome evolution in the convergent and complex bioluminescent organs of squid. *Proceedings of the National Academy of Sciences of the United States of America* 111, E4736–4742.

Rifkin, S.A., Kim, J., and White, K.P. (2003) Evolution of gene expression in the Drosophila melanogaster subgroup. *Nature Genetics* 33, 138–144.

Robinson, M.D., McCarthy, D.J., and Smyth, G.K. (2010) edgeR: A Bioconductor package for differential expression analysis of digital gene expression data. *Bioinformatics* 26, 139–140.

Rohlfs, R.V., Harrigan, P. and Nielsen, R. (2014) Modeling gene expression evolution with an extended Ornstein-Uhlenbeck process accounting for within-species variation. *Molecular Biology and Evolution* 31, 201–211.

Ruan, H., Su, Z., and Gu, X. (2016) TreeExp1.0: R Package for Analyzing Expression Evolution Based on RNA-Seq Data. *Journal of Experimental Zoology. Part B, Molecular and Developmental Evolution* 326, 394–402.

Schraiber, J.G., Mostovoy, Y., Hsu, T.Y. and Brem, R.B. (2013) Inferring evolutionary histories of pathway regulation from transcriptional profiling data. *PLoS Computational Biology* 9, e1003255.

Sudmant, P.H., Alexis, M.S. and Burge, C. B. (2015) Meta-analysis of RNA-seq expression data across species, tissues and studies. *Genome Biology* 16, 287.

8

Ancestral Transcriptome Inference

Transcriptional gene network underlies developmental programs and evolutionary adaptation (King and Wilson 1975; Wray 2007). Ancestral state reconstruction has been shown to be a powerful methodology in the study of transcriptome evolution. In a phylogeny with diverse organisms, one can trace the evolutionary changes of gene regulation if the ancestral patterns of the transcriptome state can be inferred (Villar et al. 2014). Gu (2004) developed an empirical Bayesian procedure to infer ancestral expression inference which was based on the Brownian model. It has been shown that gene expression is generally under stabilizing selection to maintain the optimal value during the course of evolution. In this case, the Ornstein-Uhlenback (OU) model is more appropriate, as discussed in Chapter 7. Yang et al. (2020) updated the method to the OU model and implemented it for the RNA-seq data. In this chapter, we discuss these advances.

8.1 Stationary model of transcriptome evolution

8.1.1 The Ornstein-Uhlenbeck (OU) model in a phylogeny

The phylogeny designed for comparative transcriptome analysis has two parts. The first part is the conventional species tree with a specified root (O). In our study this species tree is supposed to be known (biological tree) or can be reliably inferred from the sequence data (molecular tree). The second part is the evolutionary lineage from the origin of the tissue (node Z) to the root (node O) of the species tree, with τ evolutionary time units.

For any number (n) of species with a rooted phylogeny, it has been shown that the joint density of $\mathbf{x} = (x_1, \ldots, x_n)$ follows a multi-variate normal distribution, $N(x_1, \ldots, x_n; \boldsymbol{\mu}, \mathbf{V})$, with the mean vector $\boldsymbol{\mu}$ and the variance-covariance matrix $\mathbf{V}$; the density is given by

Statistical Analysis of Molecular and Genomic Evolution. Xun Gu, Oxford University Press. © Xun Gu (2024).
DOI: 10.1093/oso/9780198816515.003.0008

$$P(\mathbf{x}) = \frac{1}{(\sqrt{2\pi})^n |\mathbf{V}|^{1/2}} \exp\left\{ -\frac{(\mathbf{x} - \boldsymbol{\mu})' \mathbf{V}^{-1} (\mathbf{x} - \boldsymbol{\mu})}{2} \right\} \tag{8.1}$$

8.1.2   Stationary Ornstein-Uhlenbeck (sOU) model

In practical analysis, some computational inconveniences may arise due to the root-dependent structure of the variance-covariance matrix ($\mathbf{V}$). Besides, the general OU model may involve too many unknown parameters that makes technically intractable in statistics. Nevertheless, we found that the stationary OU model (sOU) can considerably simplify the computational complexity. The sOU model invokes two main assumptions:

- Let z_0 be the expression level at the evolutionary time (node Z) of the emergence of a particular tissue. Since then, the evolutionary change of the expression variable (x) along the lineage follows an OU model denoted by $OU(x|z_0; , \tau, \beta, W)$, resulting a normal distribution of Eq. (8.2) at the root (O) of the species phylogeny, with the mean and the variance given by

$$\bar{x}_0 = \mu + (z - \mu)e^{-\beta\tau}$$

$$V_0 = \frac{1 - e^{-2\beta\tau}}{W} \tag{8.2}$$

We further assume that the origin of the tissue was much more ancient than the species tree under study. Consequently, the mean and the variance at the root of phylogeny satisfy $\bar{x}_0 \to \mu$ and $V_0 \to 1/W$ as $\tau \to \infty$, respectively.

- Since then, the optimal level (μ) and the strength of stabilizing selection (W) remain constant during the course of evolution along the species phylogeny. Consequently, the expression variances in all internal and external nodes are the same, which equals to $1/W$.

8.1.3   Variance-covariance matrix under sOU

An important property of the sOU model is that the variance-covariance matrix $\mathbf{V}$ is root-independent, which can be written in terms of *expression branch lengths*. Let b_k be the k-th expression branch length. Then, the covariance matrix $\mathbf{V}$ can be written as follows

$$V_{ij} = \begin{cases} 1/W & \text{if} \quad i = j \\ e^{-\sum_{k_{ij}} b_k}/W = e^{-D_{ij}}/W & \text{if} \quad i \neq j \end{cases} \tag{8.3}$$

where the subscript notation k_{ij} runs over all branches connecting between x_i and x_j, and $D_{ij} = \sum_{k_{ij}} b_k$ is the expression distance between species i and j.

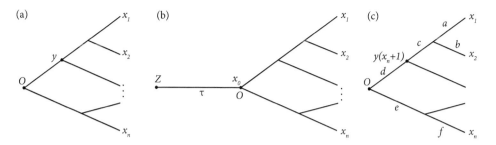

Figure 8.1 *The schematic of ancestral expression inference along a phylogeny. (A) The simplest case of n genes for empirical Bayesian method. (B) Phylogeny when considering the origin of the tissue (node Z) to the root (node O) of the species tree. (C) The case of n+1 genes for ancestral inference based on empirical Bayesian method.*

Reproduced with permission from Yang, Ruan, Zou, Su, Gu, 'Ancestral transcriptome inference based on RNA-Seq and ChIP-seq data', Methods, 176, 2020, pp. 99–105, Elsevier

8.2 Single-node ancestral transcriptome inference

8.2.1 Three-species case: an illustration

Some notations

This method provides a fast Bayesian procedure to infer ancestral expression profile because each time it deals with one ancestral node and then runs over the tree. For illustration, we start from an sOU model under a three-species phylogeny [Fig. 8.1 (A)]. Let x_1, x_2, and x_3 be the expression levels of three orthologous genes, respectively, and $P(x_1, x_2, x_3)$ be the joint density, which can be derived as follows. Denote the expression level at the ancestral node A by x_4. Let t_k and β_k ($k = 1, 2, 3$, and 4) be the evolutionary times and the evolutionary rates in the k-th branch respectively. Given the initial value (x_0) at root O, the change of x_4 follows $OU(x_4|x_0; \beta_4 t_4, W)$ and the change of x_3 follows $OU(x_3|x_0; \beta_3 t_3, W)$. Similarly, given the ancestral level x_4, the changes of x_1 and x_2 follow $OU(x_1|x_4; \beta_1 t_1, W)$ and $OU(x_2|x_4; \beta_2 t_2, W)$, respectively. To be concise in notations, let $b_k = \beta_k t_k$ as the corresponding expression branch length.

Derivation of $P(x_1, x_2, x_3, x_4)$

According to the Markov property, we obtain the joint density

$$P(x_1, x_2, x_3, x_4|x_0) = OU(x_3|x_0; \beta_3 T, W_3)OU(x_1|x_4; \beta_1 t, W)$$
$$\times\ OU(x_2|x_4; \beta_2 t, W)OU(x_4|x_0; \beta_4(T-t), W) \qquad (8.4)$$

Let $\pi(x_0)$ be the distribution of expression level at the root (O) of the phylogeny, we have

$$P(x_1, x_2, x_3, x_4) = \int_{-\infty}^{\infty} P(x_1, x_2, x_3, x_4|x_0)\pi(x_0)dx_0 \qquad (8.5)$$

Further, the sOU model claims that $\pi(x_0)$ follows a normal distribution with mean μ and variance $1/W$. After some derivations, we show that $P(x_1, x_2, x_3, x_4)$ is a four-variate normal density $N(x_1, x_2, x_3, x_4; \boldsymbol{\mu}, \mathbf{V})$, with the mean vector $\boldsymbol{\mu} = (\mu, \mu, \mu, \mu)'$, and the variance-covariance matrix $\mathbf{V}_4$ given by

$$V_4 = \frac{1}{W}\begin{pmatrix} 1 & e^{-D_{12}} & e^{-D_{13}} & e^{-b_1} \\ e^{-D_{12}} & 1 & e^{-D_{23}} & e^{-b_2} \\ e^{-D_{13}} & e^{-D_{23}} & 1 & e^{-(b_3+b_4)} \\ e^{-b_1} & e^{-b_2} & e^{-(b_3+b_4)} & 1 \end{pmatrix} \tag{8.6}$$

where $D_{12} = b_1 + b_2$, $D_{13} = b_1 + b_4 + b_3$ and $D_{23} = b_2 + b_4 + b_3$.

Derivation of $P(x_1, x_2, x_3)$

After integrating out x_4, the unobservable ancestral expression, one can show that the joint density

$$P(x_1, x_2, x_3) = \int_{-\infty}^{\infty} P(x_1, x_2, x_3, x_4) dx_4 \tag{8.7}$$

is a three-variate normal density $N(x_1, x_2, x_3; \boldsymbol{\mu}, \mathbf{V})$, with the mean vector $\boldsymbol{\mu} = (\mu, \mu, \mu)'$, and the variance-covariance matrix $\mathbf{V}_3$ given by

$$V_3 = \frac{1}{W}\begin{pmatrix} 1 & e^{-D_{12}} & e^{-D_{13}} \\ e^{-D_{12}} & 1 & e^{-D_{23}} \\ e^{-D_{13}} & e^{-D_{23}} & 1 \end{pmatrix} \tag{8.8}$$

Derivation of $P(x_4|x_1, x_2, x_3)$

According to the Bayes rule, the posterior density $P(x_4|x_1, x_2, x_3)$ is computed as follows

$$P(x_4|x_1, x_2, x_3) = \frac{P(x_1, x_2, x_3, x_4)}{P(x_1, x_2, x_3)} \tag{8.9}$$

It has been shown that the posterior density $P(x_4|x_1, x_2, x_3)$ follows a normal distribution. Hence, the posterior mean of x_4, $E[x_4|x_1, x_2.x_3]$, is used as a Bayesian inference of the ancestral expression at node A, which is a linear function of x_1, x_2 and x_3 (see below for a general treatment).

8.2.2   General case in a phylogeny

Mathematical formulation

Let $\mathbf{x} = (x_1, \ldots, x_n)$ be the observed expression profile and y be the expression level at the ancestral node of interest. According to the Bayes rule, the posterior density $P(y|x_1, \ldots, x_n)$ is computed as follows

$$P(y|x_1,\ldots,x_n) = \frac{P(x_1,\ldots,x_n,y)}{P(x_1,\ldots,x_n)} \tag{8.10}$$

From Eq. (8.10), under the stationary OU model, it is known that $P(x_1,\ldots,x_n)$ is an n-variate normal density, with the mean vector $\boldsymbol{\mu} = (\mu,\ldots,\mu)'$ and the variance-covariance matrix $\mathbf{V}$. In the same manner, $P(x_1,\ldots,x_n,y)$ is an $n+1$-variate normal density (y is treated as the $n+1$-th element), with the mean vector $\boldsymbol{\mu} = (\mu,\ldots,\mu)'$ and the variance-covariance matrix $\mathbf{U}$. The structure of $\mathbf{U}$ is as follows: If $1 \le i,j \le n$, the ij-th element of $\mathbf{U}$ is equal to that of $\mathbf{V}$ given by Eq. (8.11). For any $i, n+1$-th element, $i = 1,\ldots,n+1$, it is given by

$$V_{i,n+1} = \begin{cases} 1/W & \text{if } i = n+1 \\ e^{-B_{iy}}/W & \text{if } i \neq n+1 \end{cases} \tag{8.11}$$

and $U_{n+1,i} = U_{i,n+1}$, where B_{iy} is the expression length of lineage from the root the i-th species to the ancestral node y.

Analytical results

Let c_{ij} be the ij-th element of $\mathbf{C} = \mathbf{U}^{-1}$. Gu (2004) has shown that the posterior density $P(y|x_1,\ldots,x_n)$ is a normal density, given by Eq. (8.10). After some algebra we obtain

$$P(y|x_1,\ldots,x_n) = \frac{1}{\sqrt{2\pi}\sigma_{y|x}} \exp\left\{ -\frac{1}{2\sigma_{y|x}^2}\left[y - \mu + \sum_{i=1}^{n} \frac{c_{i,n+1}}{c_{n+1,n+1}}(x_i - \mu) \right]^2 \right\} \tag{8.12}$$

where $\sigma_{y|x}^2 = 1/c_{n+1,n+1}$ is the (posterior) variance of y. That is, the posterior mean of y conditional of $\mathbf{x} = (x_1,\ldots,x_n)'$ is given by

$$E[y|x_1,\ldots,x_n] = \beta_0 + \sum_{i=1}^{n} \beta_i x_i \tag{8.13}$$

where $\beta_i = -c_{i,n+1}/c_{n+1,n+1}$ and $\beta_0 = \mu(1 + \sum_{i=1}^{n} \beta_i)$. Apparently, the posterior mean prediction for the ancestral gene expression is a linear function of current gene expressions.

8.2.3 Algorithm implementation

For RNA-seq data, a single expression value for each gene per species was obtained by taking the means of TPM values; this is considered as a common normalized measure that takes the sequencing depth and transcript length into account between replicates. For all samples, genes with TPM less than 1 in any species were removed and values were then log2-transformed with a pseudocount of 1 added prior to the log-transformation for the further analysis. Hence the transformed expression variable is non-negative and continuous.

A fast algorithm under the stationary OU model has been formulated (Yang et al. 2020). The algorithm to calculate the coefficients $b_0, b_1, \ldots, b_n$ under the stationary OU model can be briefly described as follows.

- As the expression variance is expected to be the same among all species, the expression variance (V_0) is the simple average of expression variances among species.

- The coefficient of correlation between the i-th and j-th external nodes, denoted by R_{ij}, is calculated by the standard approach.

- Let $\mathbf{R}$ be the matrix of coefficients of correlation. It is straightforward to calculate the variance-covariance matrix of $P(x_1, \ldots, x_n)$ by $\mathbf{V} = V_0\mathbf{R}$.

- (iv) The difficulty in calculating the variance-covariance matrix of $P(x_1, \ldots, x_n, y)$, $\mathbf{V}_M$, is how to calculate the covariance elements between the i-th external node and the $n + 1$-th (internal) node, $i = 1, \ldots, n$ [Fig. 8.1 (C)]. Under the stationary OU model, we show $V_{i,n+1} = V_0 e^{-d_i}$, where d_i is the expression branch length from the internal node (y) to the i-th external node. Yang et al. (2020) developed a simple method to estimate d_i by mapping the expression distance matrix onto the known phylogeny. As $R_{n+1,i} = R_{i,n+1}$ and $R_{n+1,n+1} = 1$, we then obtain $\mathbf{V}_M$ and its inverse matrix $\mathbf{C}$.

8.3 Joint ancestral inference

8.3.1 Joint posterior distributions

To explore the joint evolutionary pattern of expression changes along a species phylogeny, the single-node method may not be sufficient. Therefore, we develop an approach for joint ancestral expression inference. For a gene family with n member genes, there are m ancestral nodes when the phylogenetic tree is given. Let $\mathbf{x} = (x_1, \ldots, x_n)'$ and $\mathbf{y} = (y_1, \ldots, y_m)'$ be the vectors of current and ancestral expression levels, respectively; and $M = n + m$. The (extended) $M \times M$ variance-covariance matrix for $(\mathbf{y}', \mathbf{x}')$ is denoted by $\mathbf{V}_M$. We have shown that $P(\mathbf{y}, \mathbf{x})$ is an M-dimensional multi-normal density. It follows that the joint posterior density of ancestral nodes $\mathbf{y}$

$$P(\mathbf{y}|\mathbf{x}) = \frac{P(\mathbf{y}, \mathbf{x})}{P(\mathbf{x})} = \frac{N(\mathbf{y}, \mathbf{x}; \boldsymbol{\mu}, \mathbf{V}_M)}{N(\mathbf{x}; \boldsymbol{\mu}, \mathbf{V})} \qquad (8.14)$$

is also $m \times m$ multi-normal, that is, $P(\mathbf{y}|\mathbf{x}) = N(\mathbf{y}; \boldsymbol{\mu}_{y|x}, \Sigma_{y|x})$, where $\boldsymbol{\mu}_{y|x} = (\mu_{y_1|x}, \ldots, \mu_{y_m|x})'$ is the posterior mean vector of the ancestral nodes, and $\Sigma_{y|x}$ is the $m \times m$ posterior variance-covariance matrix of $y_1, \ldots, y_m$.

8.3.2   Decomposition of variance-covariance matrix

To obtain useful analytical results for numerical calculation, we partition the matrix $\mathbf{V}_M$ as follows

$$\mathbf{V}_M = \begin{bmatrix} \mathbf{A} & \mathbf{H} \\ \mathbf{H}' & \mathbf{V} \end{bmatrix} \tag{8.15}$$

where $\mathbf{H}$ and $\mathbf{A}$ are $m \times n$ and $m \times m$ matrices, respectively. The matrix $\mathbf{H}$ is the ancestral-current expression covariances and $\mathbf{A}$ is the variance-covariance matrix among ancestral nodes. For instance, in the case of double-node ancestral transcriptome inference, we have

$$\mathbf{A} = \frac{1}{W}\begin{pmatrix} 1 & e^{-D_{12}} \\ e^{-D_{12}} & 1 \end{pmatrix} \tag{8.16}$$

Thus, the inverse of the matrix $\mathbf{V}_M$ can be written as

$$\Lambda_M = \begin{bmatrix} \mathbf{A} & \mathbf{H} \\ \mathbf{H}' & \mathbf{V} \end{bmatrix}^{-1} = \begin{bmatrix} \Lambda_{yy} & \Lambda_{yx} \\ \Lambda'_{yx} & \Lambda_{xx} \end{bmatrix} \tag{8.17}$$

where Λ_{xx}, Λ_{xy}, and Λ_{yy} are $n \times n$, $m \times n$, and $m \times m$ matrices, respectively.

8.3.3   Joint ancestral inferences by posterior means

Gu (2004) has shown that the joint ancestral inferences by the means of posterior means is given by

$$\boldsymbol{\mu}_{y|x} = \boldsymbol{\mu} - \Lambda'_{yx}\Lambda_{yy}^{-1}(\mathbf{x} - \boldsymbol{\mu}) \tag{8.18}$$

It appears that the joint ancestral inferences are still the linear functions of expression levels in the current species. The stochastic variance-covariance matrix is computed by

$$\Sigma_{y|x} = \Lambda_{yy}^{-1} \tag{8.19}$$

8.4   Some technical issues

8.4.1   Effects of phylogenetic structure

The statistical method we developed requires a pre-specified coefficient of correlation matrix $\mathbf{R}$. One may ask whether the phylogenetic effects can be sufficiently accounted for. A simple way to infer the expression phylogeny is to see whether it virtually reconstructs the species tree. For instance, Brawand et al. (201) reported the RNA-seq analysis from six tissues across major mammals and birds, and showed that the expression phylogeny of each tissue largely reflected the evolutionary relatedness of these species. Our future study includes extensive simulations to determine to what extent our ancestral inference is reliable before the phylogenetic signals vanish.

8.4.2   Extension to many-to-many orthologous genes

For simplicity, our analysis only used one-to-one orthologous genes, but our method would not be restricted by this technical limitation. To deal with the case of not one-to-one orthologous, one may classify any gene in to three states: zero-state (no identified orthologous gene), one-copy state, and many-copy state. In transcriptome evolution, zero-state can be treated as an equivalence to the one-copy state without expression. Two options are available; the total expression level under the dosage model, or the medium of expression levels under the model of expression compensation. While it remains an unsolved problem how to distinguish between the dosage effect and the expression compensation, we found that our estimation of expression conservation only has marginal difference between these two options (not shown).

8.4.3   Limitations of ancestral state inference

Like many other ancestral algorithms for the ancestral state inference, such as DNA sequences or morphological characters, our method for reconstructing the transcriptome at ancestral nodes is difficult to test directly by experimentations. There are two ways to compare the efficiency and accuracy of our nearly-developed method with other tools. The first is the statistical reliability. While the empirical Bayesian procedure of ancestral transcriptome inference is straightforward, we notice two error resources that may affect our estimates. The first one is the number of species (n) that determines the stochastic variance (V_{stoc}), i.e., the uncertainty generated during the course of evolution. And the second one is the number of biological replicates (m) that determines the sampling variance (V_{samp}). Since they are virtually independent, the statistical reliability of the inferred ancestral state (y) is approximately given by

$Var(y) = V_{stoc} + V_{samp}$. Apparently, the (posterior) variance of y, i.e., $\sigma^2_{y|x} = 1/c_{n+1,n+1}$, is an appropriate measure for V_{stoc}. As the average over (m) biological replicates is commonly used in the transcriptome analysis, we have $V_{samp} = h/m$, where h is a positive constant. In short, increasing m reduces the sampling effect rather than the stochastic effect, whereas increasing n reduces the stochastic effect rather than the sampling effect. The second way is biological significance. For instance, the evolutionary trajectory of gene expression along the phylogeny is highly correlated with a certain cellular feature, providing some predictions for the phenotype consequence of knockout or overexpression which can be tested by experiments. We expect our method will be helpful for such research.

References

Brawand, D., Soumillon, M., Necsulea, A., et al. (2011) The evolution of gene expression levels in mammalian organs. *Nature* 478, 343–348.

Gu, X. (2004) Statistical framework for phylogenomic analysis of gene family expression profiles. *Genetics*. 167, 531–542.

King, M.C. and Wilson, A.C. (1975) Evolution at two levels in humans and chimpanzees. *Science* 188, 107–116.

Villar, D., Flicek, P., and Odom, D.T. (2014) Evolution of transcription factor binding in metazoans - mechanisms and functional implications. *Nature Review Genetics* 15, 221–233.

Wray, G.A. (2007) The evolutionary significance of cis-regulatory mutations. *Nature Review Genetics* 8, 206–216.

Yang J, Ruan H, Zou Y, Su Z, Gu X (2020) Ancestral transcriptome inference based on RNA-Seq and ChIP-seq data. *Methods* 176, 99–105.

Zhang, Y., Liu, T., Meyer, C.A. et al. (2008) Model-based analysis of ChIP-Seq (MACS). *Genome Biology* 9, R137.

9

Strength of Expression Conservation in Transcriptome Evolution

Substantial evidence showed that gene expression is commonly under a stabilizing selection to maintain the optimal level (Boross and Papp 2017; Dekel and Alon 2005; Deutschbauer et al. 2005; Papp et al., 2003; Park and Lehner 2013; Sopko et al. 2006). Bedford and Hartl (2009) analyzed a group of Drosophila species and found that stabilizing selection plays a major role in limiting divergence of gene-expression level. Brawand et al. (2011) used the maximum likelihood (ML) method to estimate the strength of stabilizing selection in mammalian transcriptome evolution. Warnefors and Eyre-Walker (2012) proposed a selection measure of gene expression, based on the within-species and between-species expression variances. Meanwhile, a number of studies reported that the degree of expression conservation caused by the stabilizing selection may differ considerably among genes (Tirosh et al. 2006; Zou et al. 2011; Cui et al. 2007; Boross and Papp 2017; Park and Lehner 2013). An intriguing question is how to estimate gene-specific strength of expression conservation. Due to the small number of species involved in the comparative transcriptome datasets, the task can be fulfilled by using a Bayesian approach (Gu et al. 2019), as discussed in this chapter (Table 9.1).

9.1 Stationary Ornstein-Uhlenbeck (OU) model under a phylogeny

9.1.1 Formulation of an OU model along a phylogeny

It has been well demonstrated that the Ornstein-Uhlenbeck (OU) model that takes the stabilizing selection into account is appropriate to be the basic model of transcriptome evolution (Bedford and Hartl 2009; Brawand et al. 2011; Gilad et al. 2006; Gu and Su 2007; Lemos et al. 2005; Rohlfs et al. 2014). As shown in previous chapters, given

Statistical Analysis of Molecular and Genomic Evolution. Xun Gu, Oxford University Press. © Xun Gu (2024).
DOI: 10.1093/oso/9780198816515.003.0009

Table 9.1 *Probabilistic distributions used in modeling the transcriptome evolution and interpretations*

Distribution	Interpretation
$P(\boldsymbol{x}\|x_0)$	The joint density of expressions $\boldsymbol{x} = (x_1, \ldots x_n)$ conditional of the expression level (x_0) at root O.
$P(\boldsymbol{x}\|z_0)$	The joint density of expressions $\boldsymbol{x} = (x_1, \ldots x_n)$ conditional of the expression level (z_0) at node Z.
$P(\boldsymbol{x}; \boldsymbol{\mu}, \boldsymbol{R}, W)$ or $P(\boldsymbol{x}; \boldsymbol{\Phi})$	The joint density of expressions $\boldsymbol{x} = (x_1, \ldots x_n)$ under the stationary OU model; a normal distribution with the mean vector $\boldsymbol{\mu}$ and the variance-covariance matrix $\boldsymbol{V} = \boldsymbol{R}/W$.
$P(\boldsymbol{x}\|W; \boldsymbol{\mu}, \boldsymbol{R})$ $P(\boldsymbol{x}\|W; \boldsymbol{\Phi})$	Rewrite from $P(\boldsymbol{x}; \boldsymbol{\mu}, \boldsymbol{R}, W)$ to indicate W is a random variable.
$\varphi(W; \alpha, \bar{W})$ or $\varphi(W; \boldsymbol{\Phi})$	The gamma distribution for W-variation among genes
$P(\boldsymbol{x}; \boldsymbol{\mu}, \boldsymbol{R}, \alpha, \bar{W})$ or $P(\boldsymbol{x}; \Phi, \bar{W})$	The marginal density of expressions $\boldsymbol{x} = (x_1, \ldots x_n)$ with respect to the gamma distribution $\varphi(W; \alpha, \bar{W})$
$P(Q(\boldsymbol{x}); \alpha, \bar{W})$	Rewrite from $P(\boldsymbol{x}; \boldsymbol{\mu}, \boldsymbol{R}, \alpha, \bar{W})$ after treating the quadratic function of $\boldsymbol{x}$, $Q(\boldsymbol{x}) = (\boldsymbol{x} - \boldsymbol{\mu})^T \boldsymbol{R}^{-1}(\boldsymbol{x} - \boldsymbol{\mu})$, as the variable.
$P(W\|\boldsymbol{x}; \boldsymbol{\mu}, \boldsymbol{R}, \alpha, \bar{W})$ or $P(W\|\boldsymbol{x}; \boldsymbol{\Phi})$	The posterior probability of W conditional of expressions $\boldsymbol{x} = (x_1, \ldots x_n)$.

the initial expression value x_0, the OU model predicts $x(t)$, the expression level after t evolutionary time units, which follows a normal distribution with the mean $E[x\|x_0]$ and variance $V(x\|x_0)$ given by

$$E[x\|x_0] = \mu \left(1 - e^{-\beta t}\right) + z_0 e^{-\beta t}$$

$$V(x\|x_0) = \frac{1 - e^{-2\beta t}}{W} \tag{9.1}$$

respectively, where β is the rate of expression evolution and W is the strength of expression conservation; one may write an OU process by $OU(x\|x_0; \theta)$ symbolically, where θ is the parameter vector.

Here we focus on the development of a statistical method to estimate the strength of expression conservation(W) of a gene from RNA-seq data of multiple species. In the evolutionary scenario illustrated in Fig. 9.1, the first component is the evolutionary lineage from the origin of the a particular tissue (node Z) to the root (node O) of the species phylogeny, with τ evolutionary time units, while the second component is the conventional species phylogeny with n species. Given the initial expression value z_0 at

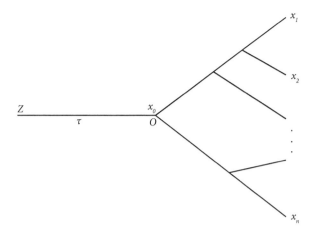

Figure 9.1 *The evolutionary scenario for comparative transcriptome analysis. Given the expression level (x_0) at root O of the species phylogeny, the joint expression density of $\boldsymbol{x} = (x_1, \ldots x_n)$ is given by $P(\boldsymbol{x}|x_0)$, which is multi-variate normal distributed. Meanwhile, the evolutionary lineage from the origin of the tissue (node Z) to the root (node O), with τ evolutionary time units. Given the initial value z_0 at node Z, the OU process of x_0 along the lineage from Z to O is given by $OU(x_0|z_0; \mu, \beta \tau, W)$. When the tissue origin is so ancient that $\tau \to \infty$, it is called the stationary OU model along the species phylogeny.*

node Z, the OU process of x_0 in the lineage from Z to O is given by $OU(x_0|z_0; \theta)$, where the parameter vector $\theta = (\mu, \beta\tau, W)$. The joint density of expressions $\mathbf{x} = (x_1, \ldots, x_n)$ on the condition of the expression level (x_0) at root O, denoted by $P(\mathbf{x}|x_0)$, can be derived under the OU model. Hansen and Martins (1996) showed that either $P(\mathbf{x}|x_0)$ or $P(\mathbf{x}|z_0)$ is multivariate normally distributed. It follows that the joint expression density of $\mathbf{x} = (x_1, \ldots, x_n)$ is given by

$$P(\mathbf{x}|z_0) = \int_{-\infty}^{\infty} OU(x_0|z_0; \tau)P(\mathbf{x}|x_0)dx_0 \qquad (9.2)$$

In the same manner, one can show that $P(\mathbf{x})$ is normally distributed, that is, $P(\mathbf{x}) \sim N(\mathbf{x}; \mu, \mathbf{V})$, where μ is the mean vector and $\mathbf{V}$ is the variance-covariance matrix.

9.1.2   The stationary assumption

The joint density of expressions along a phylogeny as given by Eq. (9.2) is difficult to apply in practice, because the current transcriptome data $(x_1, \ldots, x_n)$ contain little information about the evolution from node Z to node O. And calculations of μ and $\mathbf{V}$ are usually difficult because they both depend on z_0 and τ. Gu et al. (2019) proposed a stationary OU model (sOU) that overcomes the difficulty. The sOU model assumes that, at the genome level, the biological function of a tissue-specific transcriptome is conservative during the course of species evolution. Technically, the sOU model involves two assumptions.

(*i*) Origin of the tissue (node Z in Fig. 9.1) was so ancient that the evolutionary time between nodes Z and O can be approximated by $\tau \to \infty$. Consequently, the expression mean and variance at root O approach to μ and $\rho^2 = 1/W$, respectively; see Eq. (9.1).

(*ii*) The optimal expression level (μ) and the strength of expression conservation (W) remain virtually constant along the species phylogeny, that is, the expression mean and variance at all internal and external nodes are equal to μ and $1/W$, respectively.

Therefore, $P(\mathbf{x}|z_0)$ can be simplified as $P(\mathbf{x})$ that is independent of z_0, with a uniform mean vector μ, i.e., $\mu_1 = \ldots, = \mu_n$, as well as the variance-covariance matrix simply given by $\mathbf{V} = \mathbf{R}/W$, where $\mathbf{R}$ is the coefficient of correlation matrix.

Our intent is to estimate the strength of expression conservation of a gene characterized by a single parameter W. In this sense, the joint density of $\mathbf{x}$ can be symbolically written by $P(\mathbf{x}; \mu, \mathbf{R}, W)$. Together, we have

$$P(\mathbf{x}; \mu, \mathbf{R}, W) = N(\mathbf{x}; \mu, \mathbf{R}/W) \tag{9.3}$$

9.2 Variation of *W* among genes

9.2.1 The gamma distribution model of *W*

The sOU model assumes that the strength of expression conservation (W) of a gene remains a constant in species evolution but differs among genes. Substantial evidence has supported this argument (Bedford and Hartl 2009; Brawand et al. 2011; Cui et al. 2007; Park and Lehner 2013; Tirosh et al. 2006; Warnefors and Eyre-Walker 2012; Zou et al. 2011). Further, one can model W as a random variable that varies among genes according to a gamma distribution, that is,

$$\phi(W; \alpha, \bar{W}) = \frac{(\alpha/\bar{W})^\alpha}{\Gamma(\alpha)} W^{\alpha-1} e^{-\alpha W/\bar{W}} \tag{9.4}$$

where $\bar{W}$ is the mean and α is the shape parameter; a small values of α means a high degree of W-variation, and $\alpha \to \infty$ means a constant W among genes.

9.2.2   The marginal distribution of x

Rewrite the joint normal density $P(\mathbf{x}; \mu, \mathbf{R}, W)$ by $P(\mathbf{x}|W; \mu, \mathbf{R})$ to indicate that W is a random variable. From Eq. (9.3), we then have

$$P(\mathbf{x}|W; \mu, \mathbf{R}) = N(\mathbf{x}; \mu, \mathbf{R}/W) \tag{9.5}$$

To be concise, we use Θ to represent all parameters except for those specified thereafter. It follows that the marginal density of $\mathbf{x}$ is given by

$$P(\mathbf{x}; \Theta) = \int_0^\infty P(\mathbf{x}|W; \mu, \mathbf{R})\phi(W; \alpha, \bar{W})dW$$

$$= A\left(\frac{\bar{W}}{\alpha}\right)^{n/2}\left[\frac{\Gamma(n/2+\alpha)}{\Gamma(\alpha)}\right]\left[\frac{\alpha}{\alpha + Q(\mathbf{x})\bar{W}}\right]^{n/2+\alpha} \tag{9.6}$$

where $Q(\mathbf{x}) = (\mathbf{x} - \mu)^T\mathbf{R}^{-1}(\mathbf{x} - \mu)$ is a quadratic function of $\mathbf{x}$, and $A = \pi^{-n/2}|\mathbf{R}|^{-1/2}$ is a normalization constant.

9.3   An empirical Bayesian estimation of *W*

9.3.1   The posterior mean of *W* as gene-specific predictor

We adopt an empirical Bayesian procedure to predict the strength of expression conservation for a single gene. By the Bayes rule, the posterior density of W on the condition of the expression profile ($\mathbf{x}$) of a gene is given by

$$P(W|\mathbf{x}; \Theta) = \frac{\phi(W; \Theta)P(\mathbf{x}|W; \Theta)}{P(\mathbf{x}; \Theta)} \tag{9.7}$$

After some mathematical calculations, one can show that the analytical form of the posterior density of W is given by

$$P(W|\mathbf{x}; \Theta) = \frac{[\alpha/\bar{W} + Q(\mathbf{x})]^{n/2+\alpha}}{\Gamma(n/2+\alpha)} W^{n/2+\alpha-1} e^{-[\alpha/\bar{W}+Q(\mathbf{x})]W} \tag{9.8}$$

Hence, $P(W|\mathbf{x}; \mu)$ follows a gamma distribution, with the mean and variance given by

$$E[W|\mathbf{x}] = \left[\frac{\alpha + n/2}{\alpha + Q(\mathbf{x})\bar{W}}\right]\bar{W}$$

$$Var(W|\mathbf{x}) = \left[\frac{\alpha + n/2}{(\alpha + Q(\mathbf{x})\bar{W})^2}\right](\bar{W})^2 \tag{9.9}$$

respectively. Particularly, the posterior mean, $E[W|\mathbf{x}]$ can be used as the predictor for the strength of expression conservation of a gene given observed expression profile $\mathbf{x}$.

9.3.2 Relative strength of expression conservation

It has been realized that the strength of expression conservation (W_k) of gene k highly depends on the normalization method used for RNA-seq raw reads count. Hence, it is difficult to compare between two sets of estimates when they used different normalization methods. To alleviate this problem, it might be convenient to use the ratio $U_k = W_k/\bar{W}$, the relative strength of expression conservation. Suppose we have N orthologous genes under study, and the expression profile of the k-th gene is denoted by $\mathbf{x}_k, k = 1, \ldots, N$. Let $W_k = E[W|\mathbf{x}_k]$ be the posterior predictor for the strength of expression conservation of gene k. Note that the expectation of the posterior mean prediction ($E[W|\mathbf{x}]$) with respect to the marginal density $P(\mathbf{x}; \mu, \mathbf{R}, \alpha, \bar{W})$ is equal to the mean of the strength of expression conservation ($\bar{W}$), that is,

$$\int E[W|\mathbf{x}]P(\mathbf{x}; \Theta)d\mathbf{x} = \bar{W} \tag{9.10}$$

Eq. (9.10) implies that the average of U_k, the relative strength of expression conservation over all genes is roughly equal to 1, that is,

$$\frac{\sum_{k=1}^{N} U_k}{N} \approx 1 \tag{9.11}$$

The sign of approximation in Eq. (9.11) means that the sampling mean of U_k is approaching 1, because in general, the distribution of W_k is only roughly consistent with that of data distribution.

9.3.3 Statistical procedure and implementation

Suppose we have RNA-seq datasets of a particular tissue from n species, and the expression profile of each k-th gene denoted by $\mathbf{x}_k = (x_{1k}, \ldots, x_{nk})$, $k = 1, \ldots, N$. Gu et al. (2019) developed a practically feasible procedure to estimate W of each gene, which actually deals with the quadratic function of $\mathbf{x}_k$, or $Q(\mathbf{x}_k)$. The procedure is briefly described below.

Calculation of quadratic functions

There are three steps to calculate a quadratic function.

(*i*) Calculate gene-k specific mean (μ_k) by a simple average over orthologous genes, denoted by $\hat{\mu}_k$.

(*ii*) Calculate the matrix of correlation coefficients ($\mathbf{R}$) from comparative RNA-seq data by the standard approach, denoted by $\hat{\mathbf{R}}$. Thus, $\hat{R}_{ij}$ is the estimated coefficient of the expression correlation between species i and j, and $\hat{R}_{ii} = 1$.

(*iii*) Let $\mathbf{C} = \hat{\mathbf{R}}^{-1}$ and c_{ij} be the ij-th element of $\mathbf{C}$. Calculate the quadratic function of each gene k, $Q(\mathbf{x}_k)$, by

$$\hat{Q}(\mathbf{x}_k) = \sum_{i=1}^{n}\sum_{j=1}^{n} c_{ij}(x_{ik} - \hat{\mu}_k)(x_{jk} - \hat{\mu}_k) \tag{9.12}$$

where x_{ik} or x_{jk} is the expression value of gene k in species i or j, respectively.

Method of moments estimation

Because of its simplicity, Gu et al. (2019) suggested use of the method of moments (MM) to estimate α and $\bar{W}$. The first and second moments of the quadratic function $Q(\mathbf{x})$ can be calculated by the conditional expectation approach. To this end, one may recall the following mathematical result:

Mean and variance of quadratic forms

Suppose that vector $\mathbf{x}$ follows a multi-Gaussian distribution with the mean μ and the variance-covariance matrix $\mathbf{V}$. Let $\mathbf{x}'\mathbf{A}\mathbf{x}$ be any quadratic function. Then

$$E[\mathbf{x}'\mathbf{A}\mathbf{x}] = tr[\mathbf{A}\mathbf{V}] + \mu'\mathbf{A}\mu$$
$$Var[\mathbf{x}'\mathbf{A}\mathbf{x}] = 2tr[\mathbf{A}\mathbf{V}\mathbf{A}\mathbf{V}] + 4\mu'\mathbf{A}\mathbf{V}\mathbf{A}\mu \tag{9.13}$$

where $tr[.]$ is short for the trace of a matrix.

First we consider conditional expectations $E[Q|W]$ and $E[Q^2|W]$. Note that, when W is given, the vector $\mathbf{x} - \mu$ follows a multi-normal distribution with mean $\mathbf{0}$ and the variance-covariance matrix $\mathbf{R}/W$. By specifying $\mathbf{A} = \mathbf{R}$ and $\mathbf{V} = \mathbf{R}/W$ in Eq. (9.13), we have $E[Q|W] = n/W$ and $E[Q^2|W] = n/W^2 + (n/W)^2$, respectively. Applying the gamma distribution of W by Eq. (9.4), we claim

$$E[Q^k] = \int_0^\infty E[Q^k|W]\phi(W; \alpha, \bar{W})dW \tag{9.14}$$

where $k = 1, 2$, resulting in

$$E[Q] = \left(\frac{\alpha}{\alpha - 1}\right)\frac{n}{\bar{W}}$$
$$E[Q^2] = \left[\frac{\alpha^2}{(\alpha - 1)(\alpha - 2)}\right]\frac{2n + n^2}{(\bar{W})^2} \tag{9.15}$$

After replacing $E[Q]$ and $E[Q^2]$ by the sample mean and second moment of $Q(\mathbf{x}_k)$s, denoted by $\bar{Q}$ and $Var(Q)$, respectively, one can estimate α and $\bar{W}$ by

$$\hat{\alpha} = 2 + \frac{1 + n/2}{Var(Q)/(\bar{Q})^2 - (1 + n/2)}$$

$$\hat{\bar{W}} = \left(\frac{\hat{\alpha}}{\hat{\alpha} - 1}\right)\frac{n}{\bar{Q}} \tag{9.16}$$

Maximum likelihood estimation

After treating $Q(x_k)$ as the observation of gene k and rewriting Eq. (9.6), symbolically, denoted by $P(Q(\mathbf{x}_k); \alpha, \bar{W})$, one can build up an approximate likelihood function

$$L(\mathbf{X}|\alpha, \bar{W}) = \prod_{k=1}^{N} P\left(Q(\mathbf{x}_k); \alpha, \bar{W}\right) \tag{9.17}$$

and obtain the maximum likelihood estimates (MLE) of α and $\bar{W}$. Note that the MM estimates can be used as initial values for obtaining the MLEs. Moreover, the standard likelihood ratio test is applied to test the null hypothesis of no W-variation among genes, i.e., $\alpha = \infty$.

Empirical Bayesian estimation of gene-specific W

Let W_k be the empirical Bayesian estimate of the strength of expression conservation of gene k. After replacing α and $\bar{W}$ by their estimates in the first equation of Eq. (9.9), we obtain

$$W_k = \left[\frac{\hat{\alpha} + n/2}{\hat{\alpha} + Q(\mathbf{x}_k)\hat{\bar{W}}}\right]\hat{\bar{W}} \tag{9.18}$$

9.4 Examples

Gu et al. (2019) applied the newly-developed method and conducted genome-wide analyses in six tissues (brain, cerebellum, liver, kidney, heart, and testis) of mammals. The RNA-seq datasets used in this study included human, chimpanzee, gorilla, orangutan, macaque, mouse, platypus, opossum, and chicken (Brawand et al. 2011). The reads count data extracted from original paper were first processed with outliers removed and normalized to standard RPKM (Reads Per Kilobase per Million mapped reads). The median of expression values of all biological replicates of a sample was taken as the expression for that sample.

Table 9.2 *Parameter estimation for the gamma distribution model of W-variation among genes: α is the shape parameter and $\bar{W}$ is the genome-wide mean of the strength of expression conservation: All estimates are based on mammalian 5635 1:1 orthologous genes and eight species*

Tissue	α	$\bar{W}$
Brain	3.04	0.283
Cerebellum	3.17	0.247
Heart	3.55	0.195
Liver	2.85	0.272
Kidney	3.16	0.239
Testis	3.43	0.261

For the one-to-one orthologous gene set in mammals (the number of genes $N = 5635$, and the number of species $n = 8$), we estimated two model parameters (α and $\bar{W}$) for all six tissues (Table 9.2). Note that these estimates were based on the log-transformation of RPKM+1. Both estimates were roughly at the same magnitude among different tissues, and the null hypothesis of $\alpha = \infty$ (no W-variation among genes), or $\bar{W} = 0$ (no stabilizing selection) was statistically rejected (the likelihood ratio test, $P < 0.001$ in all cases).

Based on the parameters estimated from the gamma distribution (Table 9.2), we performed the empirical Bayesian inference of the strengths of expression conservations for all 5,635 1:1 orthologous genes in six tissues. The diagonal panels in Fig. 9.3 show the histograms of the estimated strength of expression conservation in six tissues (brain, cerebellum, heart, liver, kidney and testis), respectively. The overall shape is similar among tissues: with a long tail toward a high W-estimate value, most genes show an intermediate level of the strength of expression conservation. The median values of six tissues are all under 0.2, whereas there are a few genes that are under a strong strength of expression conservation (over 0.4).

We are interested whether gene expression level is correlated with the strength of expression conservation. Fig. 9.3 shows the value of W-estimates against the expression levels in the human brain and heart, respectively. In each case, we found a significant correlation between them (the $W - E$ correlation for short), but the magnitude of correlation is low, and other tissues and/or other species give virtually the same results. Tentatively, we grouped genes according to the expression level and calculate the average of W-estimate in each group. In the case of brain tissue, we found the average of W to

Table 9.3 *Top 20 genes with highest W values in mammalian brain*

W	hgnc_symbol	Description
0.5686	*THAP11*	THAP domain containing 11
0.5673	*TRAPPC12*	Trafficking protein particle complex 12
0.5549	*CTBP1*	C-terminal binding protein 1
0.5294	*MCM9*	Minichromosome maintenance 9 homologous recombination repair factor
0.5251	*STN1*	STN1, CST complex subunit
0.5175	*MFN2*	Mitofusin 2
0.5107	*OGDHL*	Oxoglutarate dehydrogenase-like
0.5078	*DOLPP1*	Dolichyldiphosphatase 1
0.5076	*MYB*	MYB proto-oncogene, transcription factor
0.5029	*KIF11*	Kinesin family member 11
0.5002	*CLCN6*	Chloride voltage-gated channel 6
0.4958	*C7orf26*	Chromosome 7 open reading frame 26
0.4957	*ABL1*	ABL proto-oncogene 1, non-receptor tyrosine kinase
0.4944	*RHOT2*	Ras homolog family member T2
0.4909	*LONP1*	Lon peptidase 1, mitochondrial
0.4879	*EPN2*	Epsin 2
0.4873	*ARFGEF2*	ADP ribosylation factor guanine nucleotide exchange factor 2
0.4837	*SPG21*	Spastic paraplegia 21 (autosomal recessive, Mast syndrome)
0.4832	*ZNF292*	Zinc finger protein 292
0.4816	*SAP130*	Sin3A associated protein 130

be 0.24 (no expression), 0.14 (low expression), 0.18 (intermedium expression) and 0.22 (high expression). We mention two interesting patterns. First, it seems counter-intuitive that genes with no expression have a high W estimate in human brain (Fig. 9.3). Further examination showed that those genes are virtually not brain-expressed in all species under study. Moreover, a similar pattern was found in the other tissues (not shown). Tentatively, we hypothesize that for those genes virtually under the "no-expression" status of a tissue, they may be subject to nontrivial purifying selection against any (gain-of-function) mutation that results in inappropriate expression, resulting in a high W value related to this tissue. Second, a positive, significant $W - E$ correlation for those expressed genes suggests that, under the regulatory network, highly expressed genes

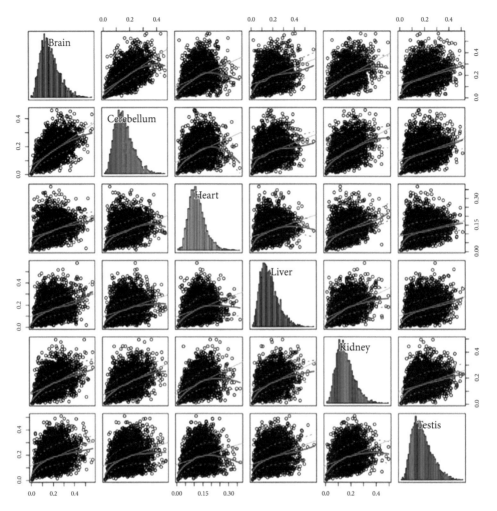

Figure 9.2 *(Diagonal panels) Histograms of the strength of expression conservation (W) of genes in six tissues (brain, cerebellum, heart, liver, kidney and testis), respectively. (Off-diagonal panels) Correlations of W between tissues. For each tissue, mammalian RNA-seq datasets of 5635 single-copy genes are used.*

may be subject to a strong stabilizing selection. However, we have to be cautious because the correlation between gene expression and strength of expression conservation is usually low. One possibility is a U-type $W - E$ correlation, i.e., genes with no-expression or genes with high expression tend to have a high W value, whereas genes with low-expression tend to have a low W value. We will further test this interesting hypothesis in our future study, as well as the relationship with DNA sequences (Tirosh and Barkai 2008).

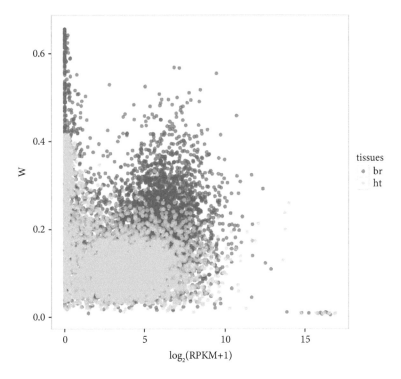

Figure 9.3 *Correlations between the strength of expression conservation and the expression level (the W-E correlation) in human brain and heart, respectively. The expression level is measured by the log-transformed value $x=log_2(RPKM+1)$.*

9.4.1 Statistical reliability of empirical Bayesian estimates

While the empirical Bayesian procedure is straightforward, there are three error resources that may affect the prediction accuracy: (i) the number of species (n) for the effect of stochastic variance (V_{stoc}), the uncertainty generated during the course of evolution; (ii) the number of biological replicates (m) for the effect of species-sampling error (V_{samp}); and (iii) the number of genes under study (N) for the gene-sampling error in parameter estimation (V_{mod}). Since they are virtually independent, the overall statistical reliability of the empirical Bayesian predictor can be measured by

$$Var(W_k) = V_{stoc} + V_{samp} + V_{mod} \tag{9.19}$$

Roughly speaking, increasing n reduces V_{stoc}, increasing m reduces V_{samp}, and increasing N reduces V_{mod}. Since N is usually much larger than the number of parameters to estimate, V_{mod} is actually negligible. Note that V_{stoc} can be computed by $V(W|x)$ according to Eq. (9.9). In comparative transcriptome analysis, a common approach is to use the expression mean over m biological replicates of the same species as input; in this case we have $V_{samp} = c/m$, where c is a positive constant.

9.5 Special topics: divergence of expression conservation between tissues

9.5.1 Covariation of W between tissues and divergence

Consider RNA-seq transcriptome datasets for two tissues (A and B), each of which includes the same species from a phylogeny. For a given orthologous gene, we try to investigate to what extent the strength of expression conservation (W) may differ between tissues. To this end, we discuss several typical patterns as follows.

(*i*) The strength of expression conservation (W) is strong in one tissue but is weak in the other one. It postulates that expression divergence between tissues results in shifted expression constraints (i.e., different strength of expression conservation) at least in some genes.

(*ii*) During the evolution along the species phylogeny, the strength of expression conservation (W) remains the same in both tissues. This pattern implies that this gene may be upregulated (highly expressed) or downregulated (lowly expressed) in both tissues under a similar pressure of stabilizing selection.

(*iii*) Expression divergence between tissues is driven by a combination of the W-divergence and the DE (differential expression) of some genes.

The popular DE-based transcriptome analysis can be directly applied to address pattern-*ii*, where orthologous genes are treated as biological replicates. Here we attempt to formulate a statistical framework for the W-divergence between tissues.

9.5.2 A two-state model of W-divergence between tissues

For any given gene, the expression profile in tissue A or B is denoted by $\mathbf{x}_A$ or $\mathbf{x}_B$, which follows a multivariate distribution denoted by $P(\mathbf{x}_A|W_A, \Theta_A)$ or $P(\mathbf{x}_B|W_B, \Theta_B)$, respectively. With respect to W_A and W_B, the strengths of expression conservation of a gene in tissue A and tissue B, respectively, all genes can be classified into either of two states:

Expression divergence-unrelated (E_0)

The strength of expression conservation (W) of an E_0-gene is the same between tissues, indicating no shift in expression constraints, that is, $W_A = W_B = W$. It follows that the joint distribution of $\mathbf{x}_A$ and $\mathbf{x}_B$ on the condition of E_0 is given by

$$P(\mathbf{x}_A, \mathbf{x}_B; \Theta | E_0) = \int P(\mathbf{x}_A | W; \Theta_A) P(\mathbf{x}_B | W; \Theta_B) \phi(W) dW \qquad (9.20)$$

Expression divergence-related (E₁)

The strength of expression conservation (W) of an E_1-gene is assumed to have no correlation between tissues, because those genes may have experienced shifted expression constraints. Consequently, W_A and W_B are assumed to be independent. In this case, the strength of expression conservation could be very strong in one tissue but very weak in the other one. Hence, the joint distribution of $\mathbf{x}_A$ and $\mathbf{x}_B$ on the condition of E_1 is given by

$$P(\mathbf{x}_A, \mathbf{x}_B; \Theta | E_1) = \int P(\mathbf{x}_A | W_A; \Theta_A) \phi(W_A) dW_A \int P(\mathbf{x}_B | W_B; \Theta_B) \phi(W_B) dW_B$$

$$= P(\mathbf{x}_A; \Theta_A) P(\mathbf{x}_B; \Theta_B) \qquad (9.21)$$

Two-state mixture model of W-divergence

In practice we do not know to which state each gene may belong. This problem can be solved by implementing a two-state probabilistic model: a gene is in the state of E_1 with a probability of $P(E_1)$, or the state of E_0 with a probability of $P(E_0)$. We denote $\tau_w = P(E_1)$, called the coefficient of expression divergence. We then have a two-state mixture model as follows

$$P(\mathbf{x}_A, \mathbf{x}_B; \Theta) = (1 - \tau_w) P(\mathbf{x}_A, \mathbf{x}_B; \Theta | E_0) + \tau_w P(\mathbf{x}_A; \Theta_A) P(\mathbf{x}_B; \Theta_B) \qquad (9.22)$$

As τ_w increases from 0 to 1, the expression divergence between tissues increases from very weak to extremely strong. Next we shall formulate a statistical test to evaluate the significance of expression divergence between two species A and B, under the null hypothesis of $\tau_w = 0$ versus the alternative $\tau_w > 0$. In the future, we address two important issues: Whether expression divergence between tissues is statistically significant? If it is the case, how can we predict genes that are largely responsible for those expression divergences?

References

Bedford, T., and Hartl, D.L. (2009) Optimization of gene expression by natural selection, *Proceedings of the National Academy of Sciences of the United States of America*, 106, 1133–1138.
Boross, G., and Papp, B. (2017) No evidence that protein noise-induced epigenetic epistasis constrains gene expression evolution. *Molecular Biology and Evolution*, 34, 380–390.

Brawand, D., Soumillon, M., Necsulea, A., et al. (2011) The evolution of gene expression levels in mammalian organs. *Nature* 478, 343–348.

Cui, Q.H., Yu, Z., Purisima, E.O., and Wang, E. (2007) MicroRNA regulation and interspecific variation of gene expression. *Trends in Genetics*, 23, 372–375.

Dekel, E., and Alon, U. (2005) Optimality and evolutionary tuning of the expression level of a protein. *Nature*, 436, 588–592.

Deutschbauer, A.M., Jaramillo, D.F., Proctor, M., et al. (2005) Mechanisms of haploinsufficiency revealed by genome-wide profiling in yeast. *Genetics*, 169, 1915–1925.

Gilad, Y., Oshlack, A., Smyth, G.K., Speed, T.P., and White, K.P. (2006) Expression profiling in primates reveals a rapid evolution of human transcription factors. *Nature*, 440, 242–245.

Gu, X., Ruan, H., and Yang, J. (2019) Estimating the strength of expression conservation from high throughput RNA-seq data. *Bioinformatics (Oxford, England)* 35, 5030–5038. doi:10.1093/bioinformatics/btz405

Gu, X., and Su, Z.X. (2007) Tissue-driven hypothesis of genomic evolution and sequence-expression correlations. *Proceedings of the National Academy of Sciences of the United States of America* 104, 2779–2784.

Hansen, T.F. and Martins, E.P. (1996) Translating between microevolutionary process and macroevolutionary patterns: The correlation structure of interspecific data, *Evolution*, 50, 1404–1417.

Lehner, B. (2013) Genotype to phenotype: lessons from model organisms for human genetics. *Nature Reviews Genetics* 14, 168–178.

Lemos, B., Meiklejohn, C.D., Caceres, M., and Hartl, D.L. (2005) Rates of divergence in gene expression profiles of primates, mice, and flies: Stabilizing selection and variability among functional categories. *Evolution*, 59, 126–137.

Papp, B., Pal, C. and Hurst, L.D. (2003) Dosage sensitivity and the evolution of gene families in yeast. *Nature*, 424, 194–197.

Park, S. and Lehner, B. (2013) Epigenetic epistatic interactions constrain the evolution of gene expression. *Molecular Systems Biology*, 9.

Rohlfs, R.V., Harrigan, P., and Nielsen, R. (2014) Modeling gene expression evolution with an extended Ornstein-Uhlenbeck Process accounting for within-species variation. *Molecular Biology and Evolution*, 31, 201–211.

Sopko, R., Huang, D., Preston, N., et al. (2006) Mapping pathways and phenotypes by systematic gene overexpression. *Mol Cell*, 21, 319–330.

Tirosh, I., and Barkai, N. (2008) Evolution of gene sequence and gene expression are not correlated in yeast, *Trends in Genetics*, 24, 109–113.

Tirosh, I., Weinberger, A., Carmi, M., and Barkai, N. (2006) A genetic signature of interspecies variations in gene expression. *Nature Genetics*, 38, 830–834.

Warnefors, M. and Eyre-Walker, A. (2012) A Selection Index for Gene Expression Evolution and Its Application to the Divergence between Humans and Chimpanzees, *Plos One*, 7.

Zou, Y.Y., Huang, W, Gu, Z, and Gu, X. (2011) Predominant gain of promoter TATA box after gene duplication associated with stress responses. *Molecular Biology and Evolution*, 28, 2893–2904.

Index